# The IMA Volumes in Mathematics and its Applications

## Volume 88



Springer
*New York*
*Berlin*
*Heidelberg*
*Barcelona*
*Budapest*
*Hong Kong*
*London*
*Milan*
*Paris*
*Santa Clara*
*Singapore*
*Tokyo*

# Institute for Mathematics and its Applications
# IMA

The **Institute for Mathematics and its Applications** was established by a grant from the National Science Foundation to the University of Minnesota in 1982. The IMA seeks to encourage the development and study of fresh mathematical concepts and questions of concern to the other sciences by bringing together mathematicians and scientists from diverse fields in an atmosphere that will stimulate discussion and collaboration.

The IMA Volumes are intended to involve the broader scientific community in this process.

Avner Friedman, Director

Robert Gulliver, Associate Director

* * * * * * * * * *

## IMA ANNUAL PROGRAMS

| | |
|---|---|
| 1982–1983 | Statistical and Continuum Approaches to Phase Transition |
| 1983–1984 | Mathematical Models for the Economics of Decentralized Resource Allocation |
| 1984–1985 | Continuum Physics and Partial Differential Equations |
| 1985–1986 | Stochastic Differential Equations and Their Applications |
| 1986–1987 | Scientific Computation |
| 1987–1988 | Applied Combinatorics |
| 1988–1989 | Nonlinear Waves |
| 1989–1990 | Dynamical Systems and Their Applications |
| 1990–1991 | Phase Transitions and Free Boundaries |
| 1991–1992 | Applied Linear Algebra |
| 1992–1993 | Control Theory and its Applications |
| 1993–1994 | Emerging Applications of Probability |
| 1994–1995 | Waves and Scattering |
| 1995–1996 | Mathematical Methods in Material Science |
| 1996–1997 | Mathematics of High Performance Computing |
| 1997–1998 | Emerging Applications of Dynamical Systems |
| 1998–1999 | Mathematics in Biology |

Continued at the back

Avner Friedman

# Mathematics in Industrial Problems

## Part 9

With 117 Illustrations

Springer

Avner Friedman
Institute for Mathematics and
its Applications
University of Minnesota
Minneapolis, MN 55455
USA

*Series Editors:*
Avner Friedman
Robert Gulliver
Institute for Mathematics and
its Applications
University of Minnesota
Minneapolis, MN 55455
USA

---

Mathematics Subject Classifications (1991): 35K55, 35K60, 35K85, 35Q35, 35Q40, 35Q60, 35Q80, 35R35, 49J40, 49K35, 49L10, 49L25, 51A20, 51E23, 60G35, 60J60, 60K40, 62A15, 62F15, 68R10, 68T10, 73B05, 78B18, 73B27, 73C50, 73F10, 73G25, 76B45, 76D05, 76R50, 76T05, 76A45, 78A10, 78A60, 80A22, 81Q05, 81Q99, 82B20, 82D60, 86A20, 86A22, 86A30, 90A15, 90A80, 90B40, 90B50, 90C06, 90D25, 90D26, 90D55

---

Library of Congress Cataloging-in-Publication Data
(Revised for Part 9)
Friedman, Avner.
Mathematics in industrial problems.
(The IMA volumes in mathematics and its applications ;
v. 16, 24, 31, 38, 49, 57, 67, 83, 88)
Includes bibliographical references and index.
1. Engineering mathematics. I. Title. II. Series:
IMA volumes in mathematics and its applications ; v. 16,
etc.
TA330.F75 1988 620′.0042 88-24909
ISBN 0-387-96860-1 (pt. 1)

Printed on acid-free paper.



Production managed by Karina Mikhli; manufacturing supervised by Jeffrey Taub.
Camera-ready copy prepared by the IMA.
Printed and bound by Braun-Brumfield, Inc., Ann Arbor, MI.
Printed in the United States of America.

9 8 7 6 5 4 3 2 1

ISBN 0-387-94945-3 Springer-Verlag New York Berlin Heidelberg SPIN 10561862

# Preface

This is the ninth volume in the series "Mathematics in Industrial Problems." The motivation for these volumes is to foster interaction between Industry and Mathematics at the "grass roots level;" that is, at the level of specific problems. These problems come from Industry: they arise from models developed by the industrial scientists in ventures directed at the manufacture of new or improved products. At the same time, these problems have the potential for mathematical challenge and novelty.

To identify such problems, I have visited industries and had discussions with their scientists. Some of the scientists have subsequently presented their problems in the IMA Seminar on Industrial Problems. The book is based on the seminar presentations and on questions raised in subsequent discussions. Each chapter is devoted to one of the talks and is self–contained. The chapters usually provide references to the mathematical literature and a list of open problems which are of interest to the industrial scientists. For some problems a partial solution is indicated briefly. The last chapter of the book contains a short description of solutions to some of the problems raised in the previous volume, as well as references to papers in which such solutions have been published.

The speakers in the Seminar on Industrial Problems have given us at the IMA hours of delight and discovery. My thanks to Frank H. Stillinger (Bell Laboratories, Lucent Industries), Bernie Rudin (IBM), Larry Carson (3M), Keith Kastella (Lockheed Martin), Bill Schneider (Ford Motor Company), David Ross (Eastman Kodak), Michael Elgersma (Honeywell Technology Center), Peter Castro (Eastman Kodak), Yitzhak Shnidman (Kodak Research Labs), Erik W. Egan (Motorola), J. Allen Cox (Honeywell Corporation), David K. Misemer (3M), Craig Poling (Lockheed Martin), Leonard Borucki (Motorola), Mahesh Morjaria (General Electric), Michael L. Oristaglio (Schlumberger-Doll Research), Giuseppe Rossi (Ford Motor Company), and Kenneth N. Morman, Jr. (Ford Motor Company).

Patricia V. Brick typed the manuscript and drew majority of the figures; she did a superb job. Thanks are also due to the IMA staff for sustaining a supportive environment. Finally, I thank Robert Gulliver, the Associate Director of the IMA, for his continual encouragement in this endeavor.

Avner Friedman
Director
Institute for Mathematics
and its Applications
July 24, 1996

# Contents

# 1

# Sphere packing problems: from the obvious to the puzzling

The general sphere packing problem is concerned with the various ways to pack a given domain with non-overlapping spheres, usually of the same size. The problem is motivated by a wide variety of potential applications in the physical and biological sciences. These include atom and ion packing in crystallography, structure and properties of amorphous materials (glasses, defective crystals), colloidal aggregates, $SiO_2$ in photonic devices, and properties of catalytic beds and of porous material. Packaging of products such as ball bearings is another possible application. Optimal transmission of digital signals can be formulated as a version of sphere packing problem (called the kissing-number problem):

A signal consists of words, each word being a sequence of $n$ bits. A bit is realized by a voltage. Since the power of transmission of a sequence of voltages is proportional to the sum of their squares, we can identify a word with a point in $\mathbb{R}^n$ and, for a transmitter with a given power, all the points lie on the boundary of a sphere. In order to avoid errors these points should be depicted in such a way that all their mutual distances are $\geq \rho$ where $\rho$ is a prescribed level (depending on the signal to noise ratio). For the most efficient transmission one must then determine the maximal number of non-overlapping balls of radius $\rho$ with centers on the surface of a given sphere. More details can be found in [1]. A similar kissing-number problem may be associated with the adsorption of molecules or virus to a surface.

On September 22, 1995 Frank H. Stillinger from AT&T Bell Laboratories described recent joint research with Lubachevsky [2][3] and with Lubachevsky and Pinson [4] on numerical and computer simulation of sphere and disk packing. The underlying feature is the random generation of sphere packing by concurrent techniques. The results they obtain are unanticipated. In 3 dimensions their sphere packing process leads to amorphous material [3]. In 2 dimensions the packed structure is crystal like, exhibiting also dislocations and vacancies [2] [3]. Small impurities (e.g., if one sphere is larger than all the others) have surprisingly significant impact on the crystal structure [4]. Based on these experimental results, Stillinger posed a variety of challenging open problems.

## 1.1 Optimal packing

We first describe lattice packing. Let $\bar{a}_1, \bar{a}_2, \ldots, \bar{a}_n$ be $n$ linearly independent points in $\mathbb{R}^n$ and introduce the lattice

$$L = \left\{ \sum_{i=1}^{n} u_i \bar{a}_i \; ; \; u_i \text{ integer} \right\} .$$

We place a ball of small radius $\varepsilon$ about each of the lattice points and then uniformly increase their radius until the balls begin to overlap. At this radius $\rho$ we then achieve sphere packing of $\mathbb{R}^n$. Denote by $\delta_R(L)$ the fraction of volume of a cube $\{-R \le x_i \le R\ ,\ 1 \le i \le n\}$ which is covered by the balls. The number $\delta(L) = \lim_{R\to\infty} \delta_R(L)$ is called the *density* of the packing corresponding to the lattice $L$, and

$$\delta = \sup_L \delta(L)$$

is called the *maximal density* of lattice packing.

In 1773 Lagrange proved that for $n = 2$ the maximal density of lattice packing is

$$\delta = \frac{\pi}{\sqrt{12}} \approx 0.9069 ,$$

and it is achieved for the lattice generated by

$$\bar{a}_1 = (2, 0),\ \bar{a}_2 = (1, \sqrt{3}) \; ;$$

this lattice is called the triangular crystal.

In 1831 Gauss proved that for $n = 3$ the maximum density for lattice packing is

$$\delta = \frac{\pi}{\sqrt{18}} \approx 0.7404 .$$

It can be visualized as follows [1]: Put three spheres on a plane so that their centers form an equilateral triangle. Continue adding spheres on the plane in such a way that each new sphere touches at least two spheres already in place. In this way we form a first layer. To construct a second layer place each new sphere in the depression created at the center of any triangular group of spheres in the first layer. The second layer is identical with the first one except for a shift in the horizontal plane. The third layer is added in the same way, etc. The lattice corresponding to this packing is given by

$$\bar{a}_1 = (2, 0, 0),\ \bar{a}_2 = (1, \sqrt{3}\ , 0), \bar{a}_3 = \left(1, \frac{3}{\sqrt{8}}\ , \frac{\sqrt{15}}{\sqrt{8}}\right) .$$

The maximal lattice packings are known for dimensions up to 8; see [5].

The situation is far less known for general packing, i.e., for packing in which the centers can be assigned at arbitrary points. For $n = 2$ the optimal packing is achieved by the lattice packing. For $n = 3$ it is strongly believed that the optimal packing is again achieved by lattice packing, but the best density established by rigorous proof so far is $\approx 0.7796$ [6].

There is also interest in sphere packings that are sparse. For $n = 2$ it is believed that the packing described in Figure 1.1 is the most sparse ($\approx 75\%$), but this has not been rigorously proved.

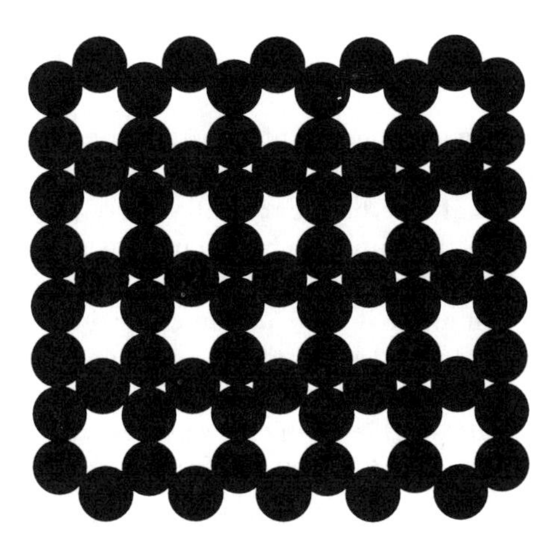

FIGURE 1.1.

## 1.2 Concurrent random packing

There are two ways to pack spheres: sequentially and concurrently. In sequential construction we add one sphere at each step so that it touches several spheres already in place; see Figure 1.2. In concurrent construction all the sphere centers are given initially, and we begin to grow them and to move them around until total jamming occurs; see Figure 1.3

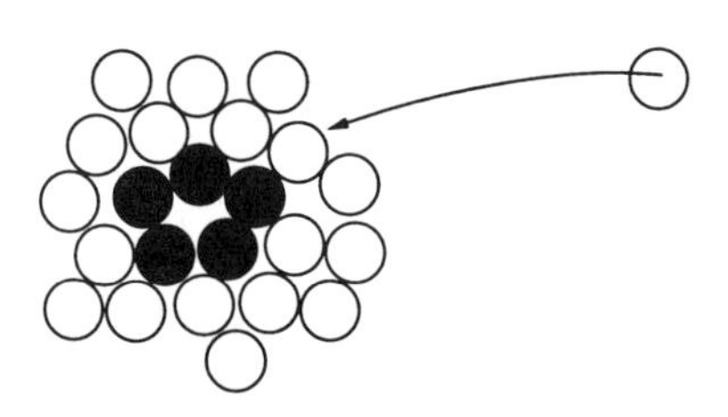

FIGURE 1.2.

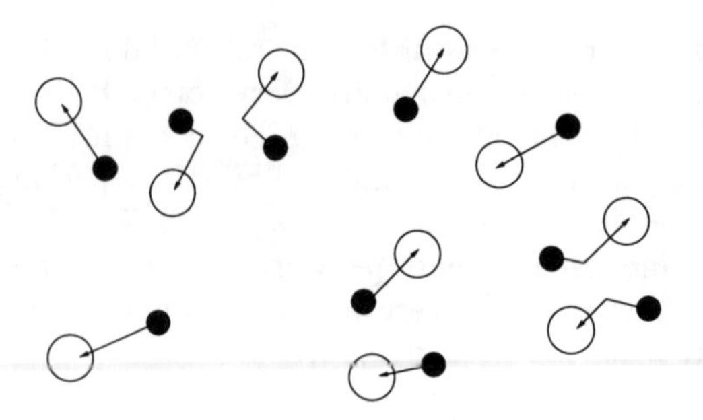

FIGURE 1.3.

We shall deal here with concurrent constructions; they lead to richer structures than sequential constructions [2–4].

If we pack a domain $\Omega$, the boundary will clearly affect the structure of the packing in the bulk. To eliminate boundary effects we shall take here $\Omega$ to be a periodic cube, that is, a cube in which opposite faces are identified. Alternately we may think of it as a periodic packing of all of space; the packing is periodic in each direction of the coordinate axes.

We shall deal here with random packing, using the recipe introduced in [2]. The packing densities achieved by this process are 0.940–0.995 of the maximal density for $n = 2$ and 0.81–0.86 of the maximal density for $n = 3$.

*The construction.* (i) $N$ points $X_i$ are chosen at random at time $t = 0$.

(ii) Random velocities $\vec{V}_i = (v_{i\alpha})$ are prescribed to the $X_i$, with $-1 \leq v_{i\alpha} \leq 1$.

(iii) Uniform growth rate is imposed, so that the radius $a(t)$ at time $t$ is $a(t) = a_0 t$ $(a_0 > 0)$.

(iv) Collision dynamics is prescribed as follows:

If two spheres collide, then the outgoing velocity components in the direction of the line $\ell$ connecting their centers is increased by $a_0$, whereas the velocities in the directions transversal to $\ell$ are not changed. Thus if their centers at collision are $\vec{r}_1$ and $\vec{r}_2$, and

$$\vec{u}_{12} = \frac{\vec{r}_1 - \vec{r}_2}{|\vec{r}_1 - \vec{r}_2|} = \vec{u}_{21},$$

and if their velocities at collision are

$$\vec{V}_1 = \vec{V}_1^{(p)} + \vec{V}_1^{(t)}, \vec{V}_2 = \vec{V}_2^{(p)} + \vec{V}_2^{(t)}$$

where "$p$" and "$t$" refer to parallel and transversal to $\vec{u}_{12}$, then their velocities just after impact are prescribed by

$$\begin{aligned} \vec{V}_1^{*} &= (\vec{V}_2^{(p)} + h\,\vec{u}_{12}) + \vec{V}_1^{(t)}, \\ \vec{V}_2^{*} &= (\vec{V}_1^{(p)} + h\,\vec{u}_{21}) + \vec{V}_2^{(t)}, \end{aligned} \tag{1.1}$$

where $h = a'(t_c) = a_0$ is the growth rate at the time $t_c$ of collision. Note that (1.1) ensures that the spheres will separate. The collision rule (1.1) is not the same as for elastic collision; in fact, one can easily check that the kinetic energy increases by $h(\vec{V}_1^{(p)} - \vec{V}_2^{(p)}) \cdot \vec{u}_{21} + h^2$ (note that $(\vec{V}_1^{(p)} - \vec{V}_2^{(p)}) \cdot \vec{u}_{21} > 0$).

(v) We continue the growth until the collision rate diverges, at a rate

$$(t_J - t)^{-\beta}$$

for some $\beta > 0$.

Figure 1.4 describes schematically the profile of the radius as a function of the initial configuration point at the moment jamming occurs. Here the $x$-axis represent the multidimensional configuration space of $(X_i, \vec{V}_i)$. The growth process is allowed as long as the radius $a$ is below the curve.

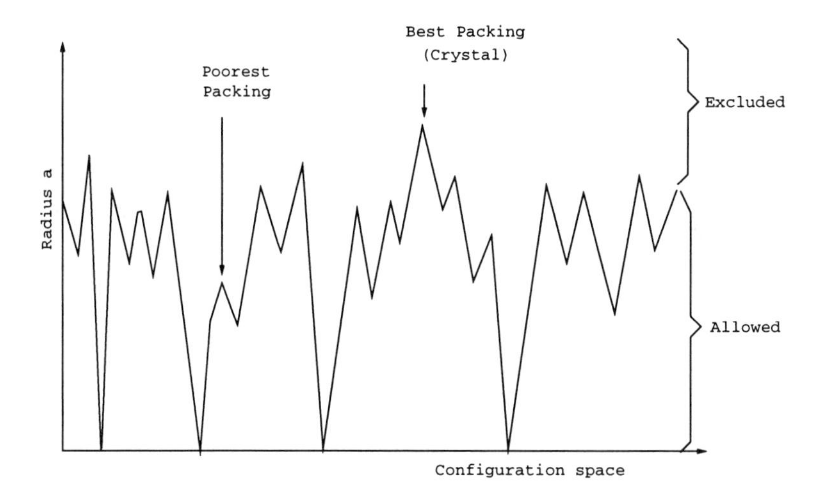


FIGURE 1.4.

## 1.3 Results for n=2 and n=3

Figure 1.5 shows a completely jammed packing formed from 27 disks in a square domain with periodic boundary conditions. Number 27 is a "rattler", i.e., it has room to move around. (If we shake the system, #27 will rattle.)

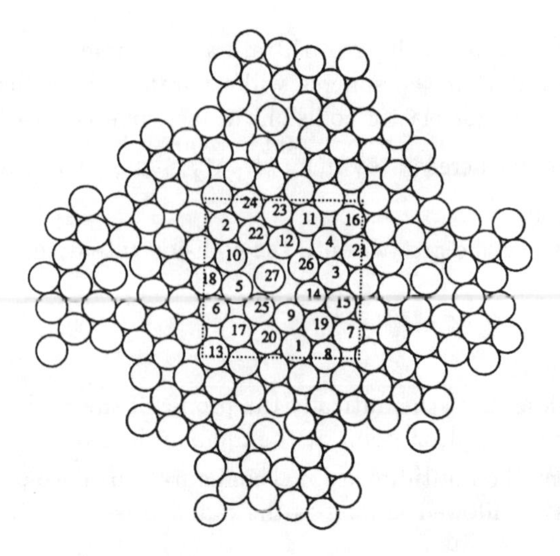


FIGURE 1.5.

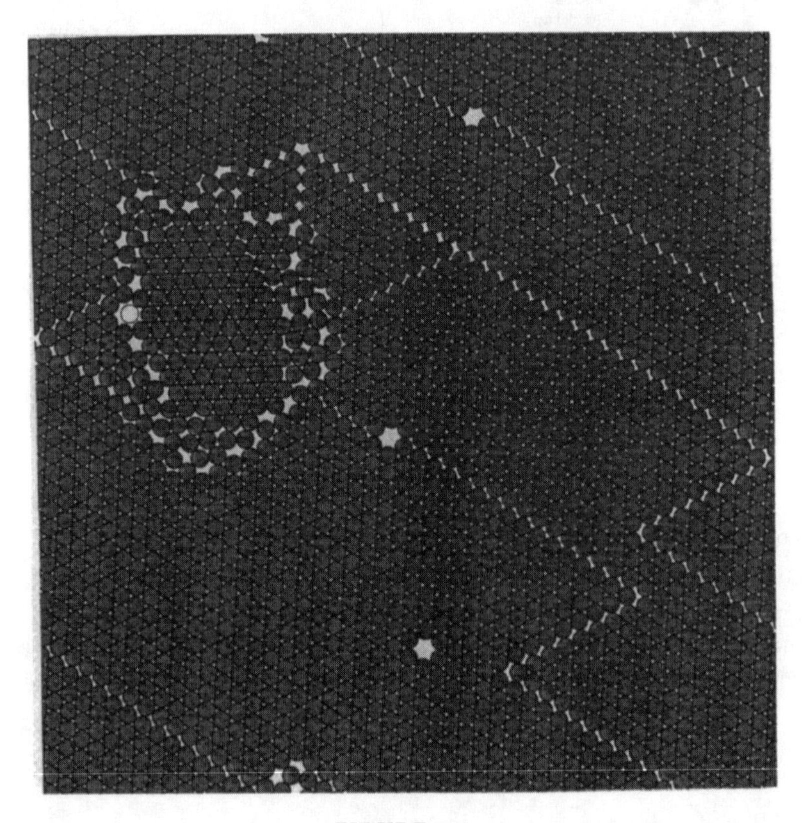

FIGURE 1.6.

Figure 1.6 is typical of the jamming features described in [2], [3] (using 2000 disks). the packing density is 0.8873. The figure reveals the presence of linear shear fractures that run across crystal grains. These fractures seem to appear only in the final states of the numerical experiments, producing sequences of nearly identical quadrilateral holes. There are three monovacancies (i.e., empty space that could accommodate one disk) and one rattler.

For $n = 3$ no substantial crystalline short-range order is present; typical irregular sphere packings are uniformly disordered. This is illustrated in Figure 1.7 which is a planar slice through an 8000-sphere irregular packing, parallel to a face of one of the periodic cells; the shaded circles represent loci of points interior to the sphere.

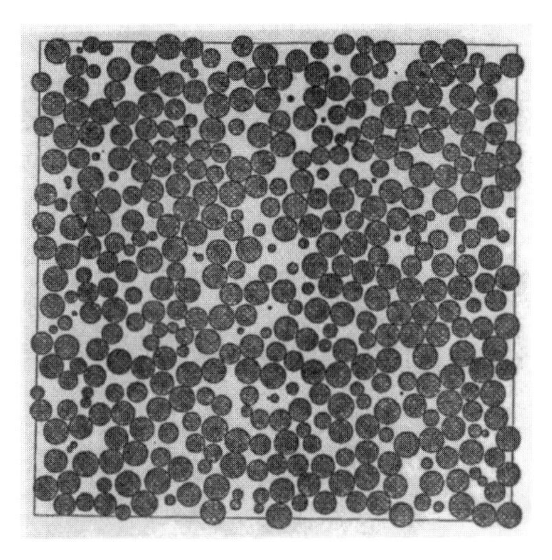

FIGURE 1.7.

We can detect significant differences between packing of disks and spheres also by looking at the following two graphs:

(i) the pair correlation function;

(ii) the histogram of number of contact numbers.

(The pair correlation function $f(r)$ counts the number of centers situated distance $r$ from each other. The contact number is the number of contact points a ball has with the other balls, i.e., the number of touching neighbors).

Figure 1.8 describes the pair correlation function for a typical disk packing. For a typical sphere packing the pair correlation function looks as in Figure 1.9

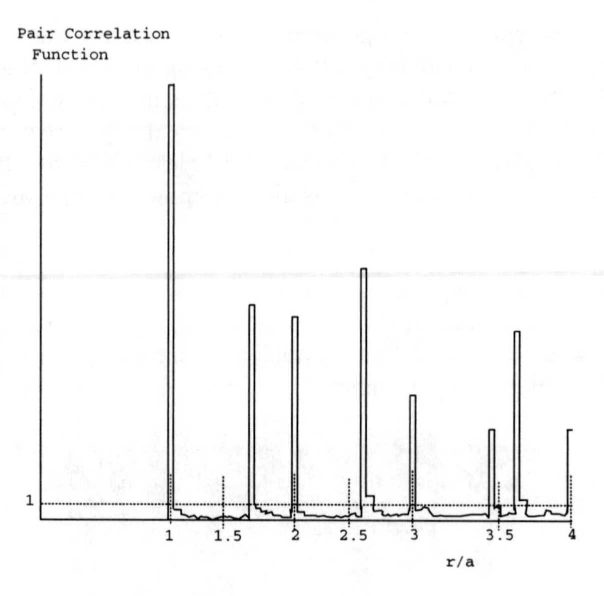


FIGURE 1.8.

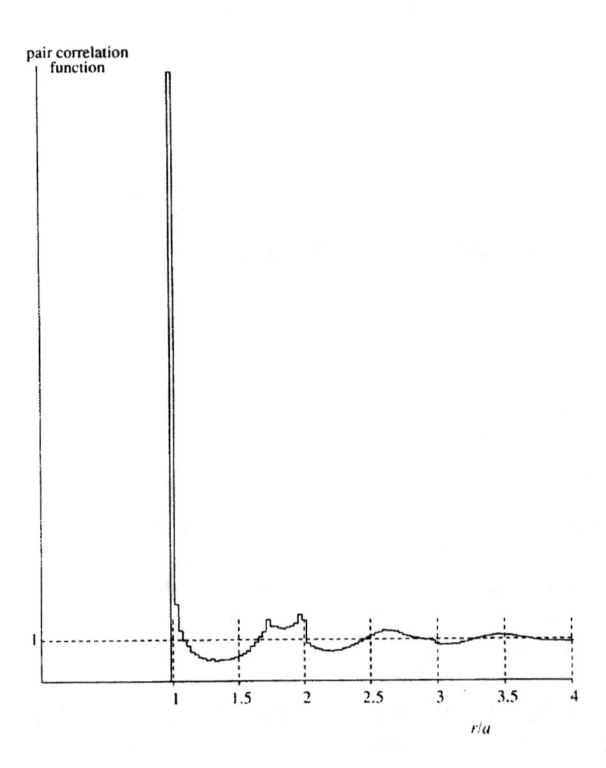


FIGURE 1.9.

Notice that in dimension 2 the pair correlation function has a spike at distance $r = a$, but it has also several additional spikes. In dimension 3 it is nearly flat beyond the first spike.

Except for rattlers, for $n = 2$ the number of contact points may vary from 2 to 6, and for $n = 3$ it may vary from 4 to 12. Figure 1.10 shows the histogram of contact numbers in a typical disk packing; note that zero and one contact points mean a rattler. Figure 1.11 shows a typical histogram for sphere packing.

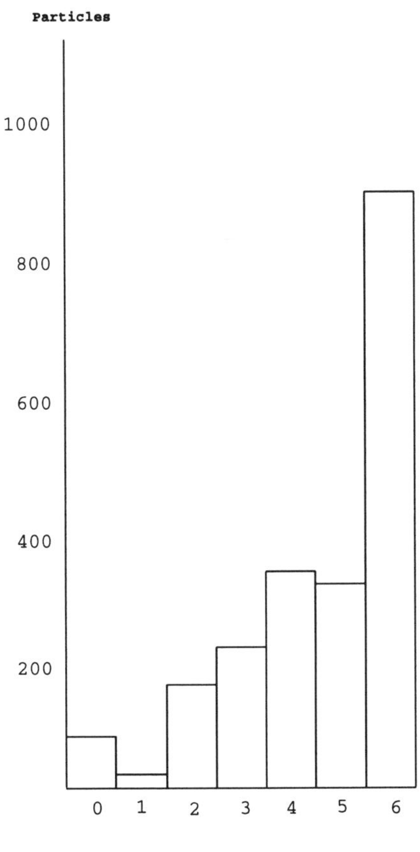


FIGURE 1.10.

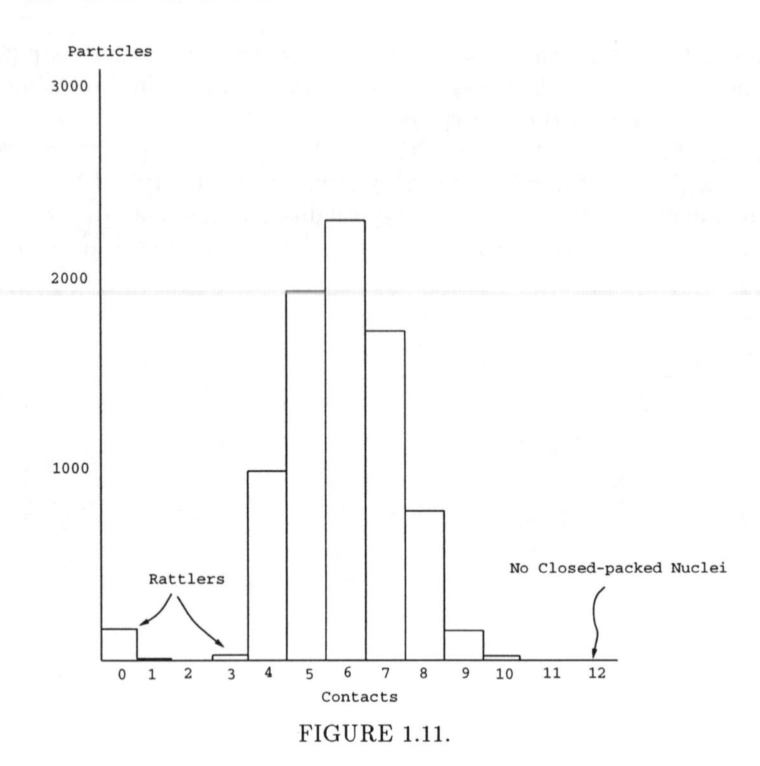


FIGURE 1.11.

Note that for $n = 3$ there are hardly any spheres with contact number 10 and none at all for contact numbers 11, 12; no close-packed nuclei! For more details on these unexpected experimental results see [2] [3]. The obvious question that arises is:

Can one rigorously prove that the above process of random packing produces the features described by the above experimental graphs?

## 1.4 Impurities

In this section we describe some results from [4] for the case where all the disks are of equal size except for one enlarged disk ("impurity" disk). The growth process is given as follows:

(i) Initially $3n^2$ small disks are arranged in one primary hexagon, so that their positions and their corresponding images form a perfectly periodic infinite triangular array; see Figure 1.12.

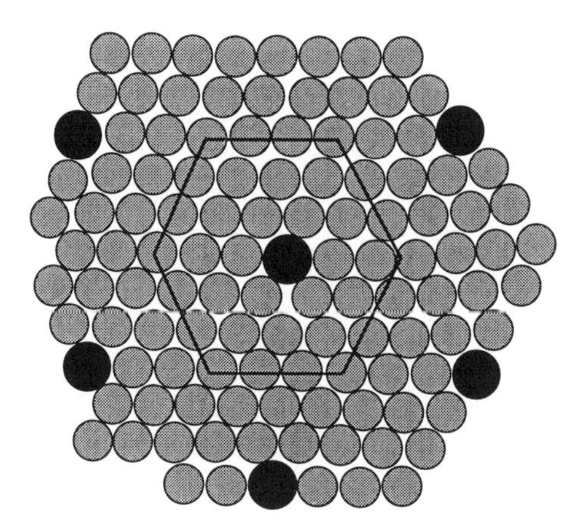

FIGURE 1.12.

(ii) All disks have the same size except the center disk (the impurity disk) whose radius is $r$ times larger than the other radii; $r$ is taken between 1 and 1.4.

(iii) Initially all the disks are separated from one another, but they cover a substantial fraction of the final packing; the impurity disk nearly overlaps with its neighboring disks.

(iv) Each disk is given initial random velocity $(v_x, v_y)$ such that

$$-\alpha \leq v_x, v_y \leq \alpha \qquad (\alpha \quad \text{constant}),$$

$$\langle v_x \rangle = \langle v_y \rangle = 0\ ,$$

$$\langle v_x^2 + v_y^2 \rangle = 1$$

where "$\langle\ \rangle$" means the average.

(v) The collision rule is the same as in §1.2. However, since near jamming the mean speed increases without bound, this is counteracted by scaling down the velocities.

For $r = 1.2$ and a very large growth rate ($a_0 = 10^4$) a typical jamming configuration looks as in Figure 1.13; it is nearly symmetric with respect to rotation by 60° (the tiny segments represent the final displacements. For $r = 1.2$ and very slow growth rate ($a_0 = 10^{-3}$), a symmetry breaking configuration develops at jamming; see Figure 1.14.

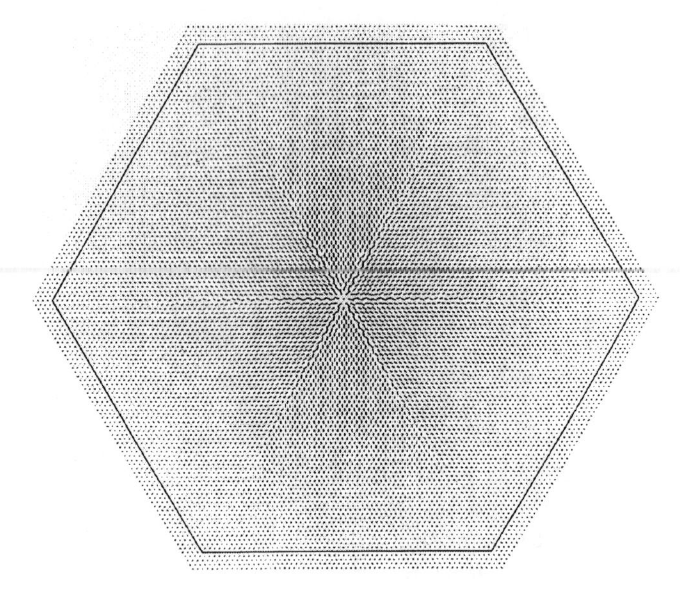

FIGURE 1.13.

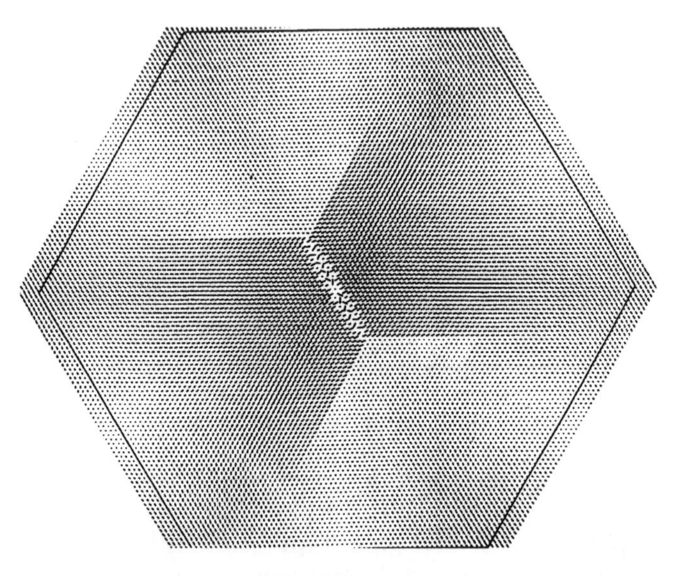

FIGURE 1.14.

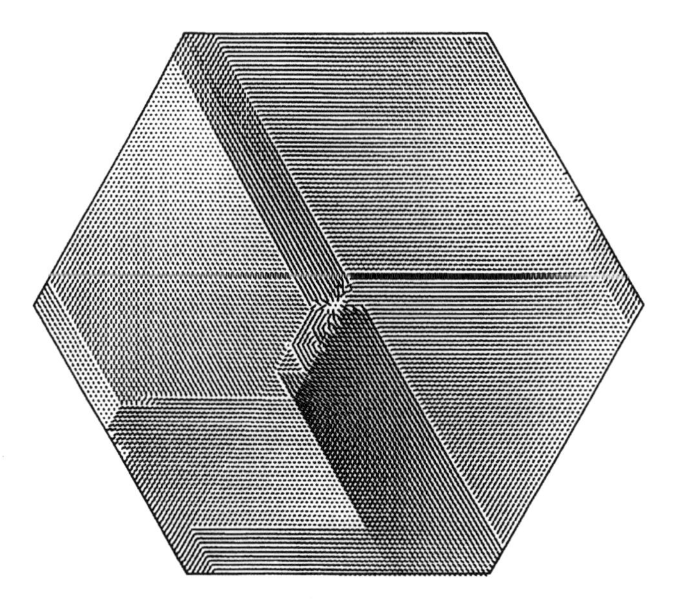

FIGURE 1.15.

By increasing $r$ (keeping $a_0 = 10^{-3}$) additional displacement patterns occur. Figure 1.15 shows a simulation result for $r = 1.4$.

The above experiments were carried out for 10,800 disks ($n = 60$). In the final jammed configurations rattler disks appear, just as in §1.3. One of the most surprising observations made in [4] is that the increase in area occupied by the disks, $\Delta A$, due to the one impurity appears to grow without bound as the system of disks (i.e., $3n^2$) increases. It is conjectured that the least upper bound on $\Delta A$ grows linearly with $n$.

## 1.5 Open problems

Some of the open problems that emerge from the previous sections are:

*Problem (1).* What is the structure of lowest-density packing for dimensions $n = 2$ and $n = 3$; are they periodic?

*Problem (2).* Given $0 < \theta < 1$, estimate the probability of achieving (with the construction of §1.2) packing density $\geq \theta$.

*Problem (3).* What is the maximum density of rattlers?

*Problem (4).* Can packing contain arbitrarily large cavities?

*Problem (5).* Estimate the contact number as a function of the packing density.

*Problem (6).* In the case of one oversized impurity, estimate the asymptotic increase of the area (for dimension 2) and of the volume (for dimension 3).

## 1.6 References

[1] N.J.A. Sloane, *The packing of spheres*, Scientific American, 250 (Jan. 1984), 116–125.

[2] B.D. Lubachevsky and F.H. Stillinger, *Geometric properties of random disk packings*, J. Statistical Physics, 60 (1990), 561–583.

[3] B.D. Lubachevsky, F.H. Stillinger, and E.N. Pinson, *Disks vs. spheres: contrasting properties of random packings*, J. Statistical Physics, 64 (1991), 501–524.

[4] F.H. Stillinger and B.D. Lubachevsky, *Pattern of broken symmetry in the impurity-perturbed rigid-disk crystal*, J. Statistical Physics, 78 (1995), 1011–1026.

[5] C.A. Rogers, *Packing and Covering*, Cambridge University Press, Cambridge (1964).

[6] C.A. Rogers, *The packing of equal spheres*, Proc. London Math. Soc., 8 (3) (1958), 609–620.

# 2

# The value of performance improvements at constant price/performance

Computer design engineers are often faced with the problem of how to justify the improvement of the performance of an apparently seldom used computer instruction even though the improvement could be substantial. The question the engineer is typically asked is, "How many more systems will we sell if we do the improvement?"

On October 16, 1995 Bernie Rudin from IBM in Austin, Texas described a model that he developed, which allows the estimation of the value of the proposed performance improvement at the price/performance ratio that has been fixed for the system being designed. Following a methodology used by business volume forecasters that relates price/performance changes to volume changes, he is then able to use the information from his value-added model to get an answer to the question posed to the engineer. His model can be extended also to other areas, such as determining whether a software purchase is advisable. He finally discussed some generalizations of his model.

## 2.1 Marginal value of instruction speedup

Most computers are designed from an instruction mix, which incorporates the experience from all designs that were done in the past. A system of instruction mix is defined by denoting each instruction by $i = 1, \ldots, n$ and setting

$$\begin{aligned} e_i &= \text{average execution time, and} \\ f_i &= \text{frequency of occurrence.} \end{aligned}$$

The frequencies must satisfy

$$\sum_{i=1}^{n} f_i = 1$$

so that the average execution time is given by

$$a = \sum_{i=1}^{n} f_i e_i \ .$$

For definiteness we shall measure time in microseconds. Then, the *performance rate* of the machine under consideration is defined as

$$m = \frac{1}{a}$$

with units in millions of instructions per second (mips). If the price, say in dollars, of the system is denoted by $p$, then the price/performance of the system is $p/m$ dollars/mips.

Suppose we redesign the instruction $j$ so that the benchmark performance is improved by a factor $s_j$. Then $s_j$ is called the *speedup* factor and the new execution time is $e_j/s_j$. Consequently, the new average execution time is given by

$$a' = \sum_{i=1}^{j-1} f_i e_i + f_j \frac{e_j}{s_j} + \sum_{i=j+1}^{n} f_i e_i \ ,$$

or

$$a' = a - f_j e_j \left(1 - \frac{1}{s_j}\right) .$$

Thus

$$\frac{a'}{a} = 1 - \frac{fe}{a}\left(1 - \frac{1}{s}\right)$$

where we have written, for brevity, $f = f_j \ , \ e = e_j \ , \ s = s_j$. The new performance is $m' = 1/a'$. We then expect a changed price $p'$. With the improved performance we expect the price to change from $p$ to $p'$, and we shall then have a new price/performance, given by $p'/m'$.

The *marginal price increase* is defined as the incremental percentage price increase,

$$\delta = 100\left(\frac{p'}{p} - 1\right) .$$

Our objective is to determine the marginal price increase associated with the performance improvement at constant price/performance, that is, when

$$\frac{p'}{m'} = \frac{p}{m} .$$

Clearly then

$$\frac{p'}{p} = \frac{m'}{m} = \frac{a}{a'}$$

so that

$$\delta = \frac{100\frac{fe}{a}\left(1 - \frac{1}{s}\right)}{1 - \frac{fe}{a}\left(1 - \frac{1}{s}\right)} . \tag{2.1}$$

We consider as an example the "divide" instruction. The "divide" time takes much longer than addition or multiplication because the designers feel that this instruction is not used frequently enough to warrant speedup. (Although if the execution time were designed to take less time, this might have caused it to be more frequently used.) In most machines the parameters are typically

$$a = 1.5 \text{ cycles}, \; e = 30 \text{ cycles}, \;\; f = 0.0005 \, . \tag{2.2}$$

Figure 2.1 describes the corresponding graph of $\delta = \delta(s)$. We note that

$$\text{if } s = 5 \text{ then } \delta = 0.81 \text{ percent;}$$

if we further increase the speedup we do not gain much improvement in the marginal price increase: indeed,

$$\text{if } s \to \infty \text{ then } \delta \to 1.01 \text{ percent}$$

On the other hand if the frequency in (2.2) is increased, $\delta$ increases more significantly; for example, if $f = 0.005$ then, for speedup of $s = 5$ , $\delta$ grows to 8.7 percent.

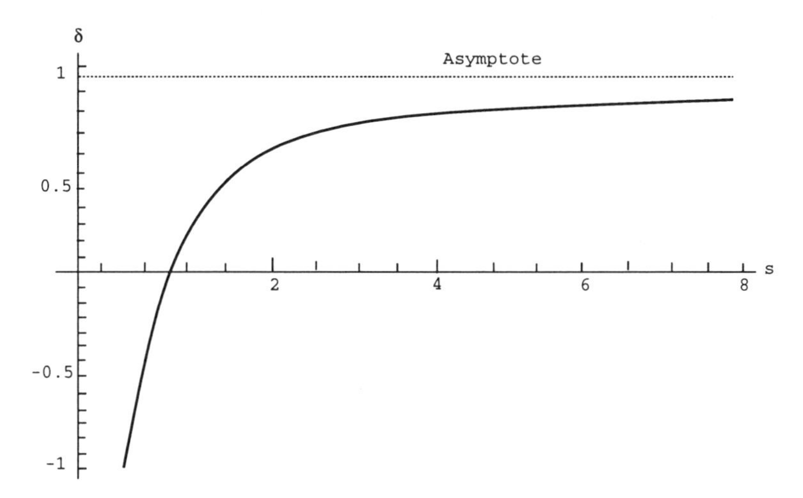


FIGURE 2.1.

## 2.2 A marginal cost model

Component development cost is a function $f(s)$ of speedup $s$. We shall assume here that

$$C' = C(1 + f(s))$$

where $C$ denotes the original system cost and $C'$ denotes the system cost after component speedup. The function $f(s)$ is a monotone increasing function such that $f(s) = 0$ when $s = 1$. We shall consider here two simple models: a linear model

$$f(s) = k(s-1)\ , \tag{2.3}$$

and a quadratic model

$$f(s) = k(s-1)^2 \tag{2.4}$$

where $k$ is a constant.

The *marginal cost* $\gamma$ of the component development is expressed as a percentage of the system cost,

$$\gamma = 100\left(\frac{C'}{C} - 1\right)\ ,$$

or

$$\gamma = 100 f(s)\ . \tag{2.5}$$

Observe that the marginal price increase is smaller (larger) than the marginal cost of the component development if $\delta(s) < \gamma(s)\quad(\delta(s) > \gamma(s))$. The break-even occurs when $\delta(s) = \gamma(s)$, or (by (2.1), (2.5))

$$f(s) = \frac{\dfrac{fe}{a}\left(1 - \dfrac{1}{s}\right)}{1 - \dfrac{fe}{a}\left(1 - \dfrac{1}{s}\right)}\ .$$

For the linear case (2.3) this occurs for

$$s = \frac{\alpha}{1-\alpha}\left(\frac{1}{k} - 1\right)\ , \quad \text{where} \quad \alpha = \frac{fe}{a}\ .$$

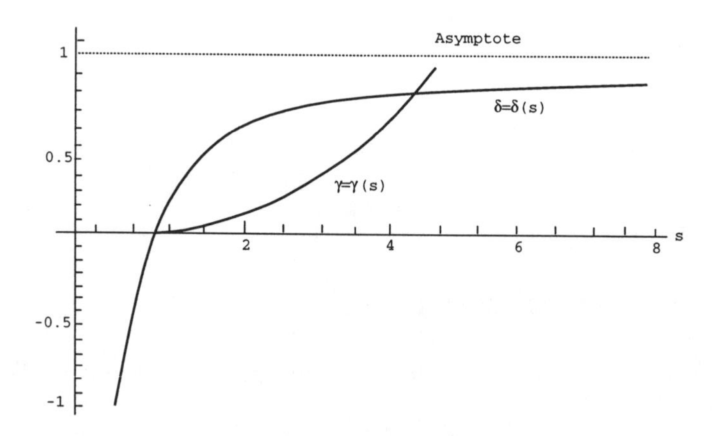


FIGURE 2.2.

Figure 2.2 describes both $\delta = \delta(s)$ and $\gamma = \gamma(s)$ in the quadratic case (2.4. The break-even speedup occurs at

$$s = \frac{1-2\alpha}{2(1-\alpha)}\left[1 \pm \left(1 + 4\frac{\alpha(1-\alpha)(1+1/k)}{(1-2\alpha)^2}\right)^{1/2}\right] .$$

## 2.3 Generalizations

Using the same notation as above, consider the effect of speedup in two components, $i = 1$ and $i = 2$. Let the corresponding speedup divisors be $s_1$ and $s_2$. Applying these speedups yields a reduced averaged instruction time

$$a' = \sum_{i=3}^{n} f_i e_i + \frac{f_1 e_1}{s_1} + \frac{f_2 e_2}{s_2} ,$$

or

$$a' = a - f_1 e_1 \left(1 - \frac{1}{s_1}\right) - f_2 e_2 \left(1 - \frac{1}{s_2}\right) ,$$

and the marginal price increase (at constant price/performance) is

$$\delta = \frac{100\left[\frac{f_1 e_1}{a}\left(1 - \frac{1}{s_1}\right)\right] + \frac{f_2 e_2}{a}\left(1 - \frac{1}{s_2}\right)}{1 - \frac{f_1 e_1}{a}\left(1 - \frac{1}{s_1}\right) - \frac{f_2 e_2}{a}\left(1 - \frac{1}{s_2}\right)} .$$

Consider the example

$$a = 1 \text{ cycle}, \; e_1 = 5 \text{ cycles}, \; e_2 = 15 \text{ cycles},$$
$$f_1 = 0.005 \; , \; f_2 = 0.003 \; .$$

Figure 2.3 describes the $(s_1, s_2)$-level curves for various values of $\delta$. Trade-offs can be studied via contour plot.

The above model for instruction improvement can be extended to other areas. We consider here briefly a model for software speedup. We denote by $i$ the $i$-th job in a job mix, with job durations (in seconds) and frequencies denoted by $E_i$ and $F_i$, respectively. Then the average job execution time is

$$A = \sum_{i=1}^{n} F_i E_i \qquad \text{(in seconds)} ,$$

and the performance is $J = 1/A$. With speedup $s$ as before we find that the average job execution $A'$ after performance improvement satisfies

$$\frac{A'}{A} = 1 - \frac{FE}{A}\left(1 - \frac{1}{s}\right)$$

where $E$ and $F$ denote the execution time and frequency of the job to which we apply the speedup. Assuming constant price/performance, that is

$$\frac{p'}{J'} = \frac{p}{J}$$

where $p$ is the price of the system and $p', J'$ are the new $p, J$ after performance improvement, we find that the marginal price increase $\Delta$ is given by (cf. (2.1))

$$\Delta = \frac{100\frac{FE}{A}\left(1 - \frac{1}{s}\right)}{1 - \frac{FE}{A}\left(1 - \frac{1}{s}\right)}$$

An example of a "long job:"

$$A = 1\,\text{min}\ ,\ E = 100\,\text{min}\ ,\ F = 0.001\ .$$

If $s = 1.25$ then the resulting marginal price increase is $\Delta = 2.04$ percent.

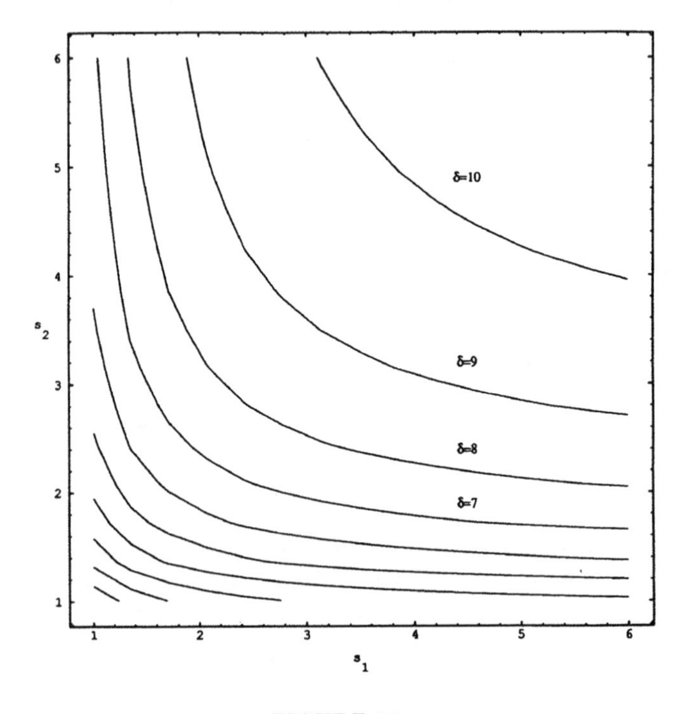


FIGURE 2.3.

If in the above example the frequency is changed to $F = 0.01$ then with the same speedup $s = 1.25$ we achieve $\Delta = 25$ percent.

The above models can be used as marketing tools to quantify added value and to evaluate options. However here one needs to combine such models with projections of the effect of price change on the sale volume. Forecasting techniques are described, for example, in [1][2].

## 2.4 References

[1] M.M. Cetron and C.A. Ralph, *Industrial Applications of Technological Forecasting*, Wiley-Interscience, New York (1971).

[2] D.G. Bails and L.C. Peppers, *Business Fluctuations, Forecasting Techniques and Applications*, Prentice-Hall, Englewood Cliffs, N.J. (1982).

# 3

# A diffusion model of droplet absorption

Many industrial applications are concerned with absorption of liquid in porous medium. One example is the coating of a film, where it is important to reduce the drying time. Another example is chemical blood tests where a blood drop is pumped out of a syringe onto a slide, where it is bound to some chemicals in dye-forming reaction and then spreads in a porous layer [1, Chap 11]. In these industrial applications the goal is to produce materials that absorb liquid in optimal fashion, e.g., minimizing the drying time under cost constraint. On November 3, 1995 Larry Carson from 3M discussed the general problem of drop absorption in a porous medium, as illustrated in Figure 3.1 (a)–(c). Experiments can measure droplet volume (above the substrate) as a function of time. Carson's goal is to develop a model that agrees with the experimental measurements.

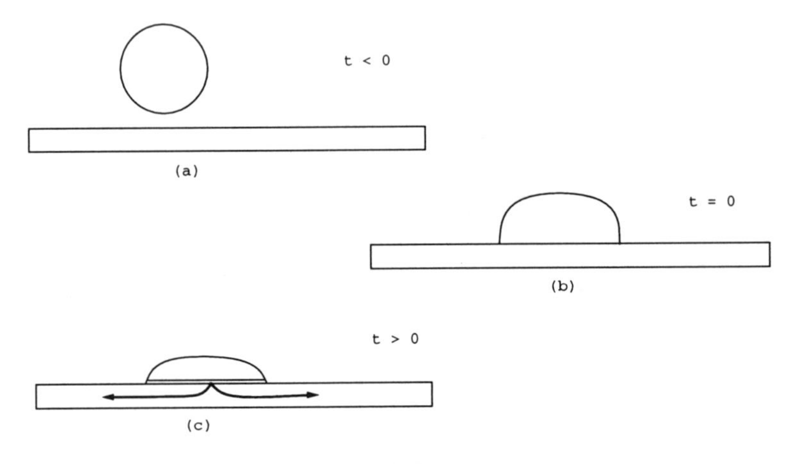


FIGURE 3.1.

## 3.1 Capillary-driven flow

Consider hydrophilic liquid in a capillary tube. Because of pressure difference across the meniscus boundary $S$ (see Figure 3.2), the liquid column $L(t)$ will continuously grow. Theory due to Lukas [2] and Washburn [3] asserts that

$$L(t) = \sqrt{D_c t} \quad \text{where} \quad D_c = (\gamma d \cos\theta)/(4\eta) \tag{3.1}$$

where $\gamma$ = surface tension, $\theta$ = contact angle, $\eta$ = liquid viscosity. Typically

$$\gamma \sim 50 dyne/cm \ , \quad \eta = 2 - 8cP$$

($cP$ = centipoise, unit viscosity in $CGS$ units). For $d \sim 10nm$ , $D_c \sim 10^{-4}\ cm^2/s$. Formula (3.1) can be verified experimentally.

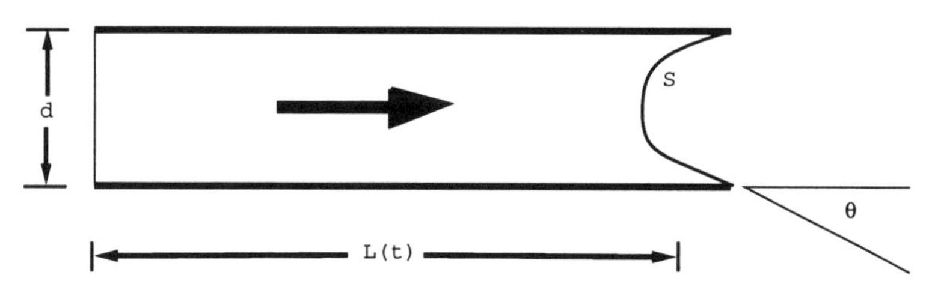


FIGURE 3.2.

A porous medium may be viewed as a network of capillaries along which fluid can diffuse; see Figure 3.3. There is no rigorous model that establishes the effective flow law of the liquid. The most commonly used models are (i) the pressure-driven flow, based on the experimental Darcy law, and (ii) the concentration-gradient-driven flow model.

FIGURE 3.3.

In Section 3.2 we shall describe a procedure to obtain a mathematical model, under the assumption that the porous medium has a periodic structure; this model provides partial theoretical justification to Darcy's law.

In Section 3.3 we shall describe the concentration-gradient-driven flow model which Carson has explored numerically. Another model is presented in Section 3.4.

## 3.2 Homogenization

An empirical relation describing the flow of fluid through a porous medium is Darcy's law [4]

$$\vec{v} = -\frac{k}{\eta}\nabla p \tag{3.2}$$

where

$\vec{v}=$ velocity of the flow,

$k =$ permeability of the porous medium,

$\eta =$ the fluid viscosity,

$p =$ the fluid pressure (in the porous medium).

Brinkman [5] suggested the following modification:

$$\nabla p = -\frac{\eta}{k}\ \vec{v} + \eta' \nabla^2 \vec{v} \tag{3.3}$$

for some $\eta' > 0$.

Both Darcy's law and Brinkman's law can be derived from the Stokes equations if we assume that the porous medium has periodic structure. The most comprehensive results in this direction are given in Allaire [6].

Let $\mathbb{R}^3$ be divided into cubes $K_i$ by parallel planes of distance $\varepsilon$ apart. Let $T$ be an open set with smooth boundary whose closure is contained in the unit cube, and set $T_\varepsilon = \varepsilon T$. We introduce in each cube $K_i$ a domain $T_{\varepsilon i}$ obtained by translation of $T_\varepsilon$, such that the $T_{\varepsilon i}$ form a periodic array, and set $\mathbb{R}^3_\varepsilon = \mathbb{R}^3 \backslash (\bigcup_i T_{\varepsilon i})$; see Figure 3.4.

Let $\Omega$ be a bounded domain in $\mathbb{R}^3$ with smooth boundary, and set $\Omega_\varepsilon = \Omega \cap \mathbb{R}^3_\varepsilon$.

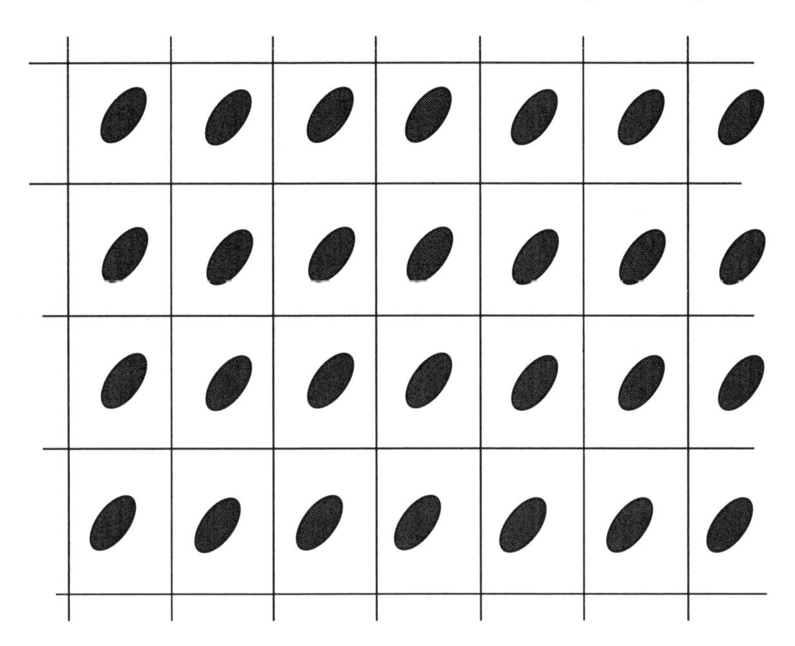

FIGURE 3.4.

Consider the Stokes equations

$$
\begin{aligned}
-\nabla^2 \vec{u}_\varepsilon + \nabla p_\varepsilon = f \quad \text{in} \quad \Omega_\varepsilon \ , \\
\nabla \cdot \vec{u}_\varepsilon = 0 \quad \text{in} \quad \Omega_\varepsilon
\end{aligned}
\tag{3.4}
$$

where $f \in L^2(\Omega)$, with the no-slip boundary condition $\vec{u}_\varepsilon = 0$ on $\partial\Omega_\varepsilon$. We can formulate the boundary condition in a weak sense by requiring that

$$\vec{u}_\varepsilon \in H_0^2(\Omega_\varepsilon) \ ;$$

we can also give (3.4) weak formulation provided $p_\varepsilon \in L^2(\Omega_\varepsilon)$. In (3.4) $p_\varepsilon$ is the pressure and $\vec{u}_\varepsilon$ is the velocity of the fluid in $\Omega_\varepsilon$.

We hope that, as $\varepsilon \to 0$,

$$\vec{u}_\varepsilon \to \vec{u} \quad \text{and} \quad p_\varepsilon \to p$$

weakly in $H_0^1(\Omega)$ and $L^2(\Omega)$, respectively, where $\vec{u}, p$ form a solution of the "homogenized" equations. This turns out to be true, but the limit depends on the asymptotic behavior of the quantities

$$\sigma_\varepsilon = \left( \frac{\varepsilon^3}{a_\varepsilon} \right)^{1/2}$$

where $a_\varepsilon$ is the volume of $T_\varepsilon$:

(i) If $\lim_{\varepsilon\to 0}\sigma_\varepsilon=\infty$ (the holes are small) than $(\vec{u},p)$ is the solution of the Stokes equations

$$-\Delta\,\vec{u}+\nabla p=f\ ,\ \nabla\cdot\vec{u}=0\quad\text{in}\quad\Omega\ ,$$

and

$$\vec{u}=0\quad\text{on}\quad\partial\Omega\ .$$

(ii) If $\lim_{\varepsilon\to 0}\sigma_\varepsilon=0$ (the holes are large) then by extending $\vec{u}_\varepsilon$ by 0 into the holes and appropriately extending also $p_\varepsilon$ into the holes, we have

$$\frac{\vec{u}_\varepsilon}{\sigma_\varepsilon}\to\vec{u}\ ,\ p_\varepsilon\to p$$

where $(\vec{u},p)$ is the solution of Darcy's law:

$$\vec{u}=M^{-1}(f-\nabla p)\ ,\ \nabla\cdot\vec{u}=0.$$

Here $M$ is a positive definite symmetric matrix which is independent of $f$; $M$ can be computed by solving the Stokes equations in $\mathbb{R}^3_\varepsilon$ with $\varepsilon=1$ for a periodic solution $(\vec{w},q)$ with $\vec{w}=0$ on $\partial T_{\varepsilon i}$ $(\varepsilon=1)$ and $\vec{w}=$ const. at $\infty$; see [6, Prop.2.1.4].

(iii) If $\lim_{\varepsilon\to 0}\sigma_\varepsilon=\sigma>0$ then $(\vec{u},p)$ is the solution of Brinkman's equations:

$$-\Delta\,\vec{u}+\nabla p+\frac{1}{\sigma^2}\,Mu=f\quad\text{in}\quad\Omega\ ,\quad\nabla\cdot\vec{u}=0\quad\text{in}\quad\Omega\ ,$$

and

$$u=0\quad\text{on}\quad\partial\Omega\ ,$$

with the same $M$ as before.

It seems difficult to infer specific properties of $M$ from specific geometric properties of the hole $T$.

The cases (i), (ii) were previously analyzed also in [7][8][9][10]. The extension of the above results to interconnected holes is given in [11]. Extensions on (i)–(iii) to slip boundary conditions are established in [12]. Finally, as mentioned in [6, p. 213], the results (i)–(iii) (with the same $M$) are valid also for the Navier–Stokes equations.

## 3.3 Concentration-gradient-driven flow

Since porous media do not generally have periodic structure, there is no absolute theoretical validation of Darcy's law (or the other laws introduced in (i), (iii) of Section 3.2). We note that under the Darcy's law, the following conditions hold at the moving boundary of the fluid boundary [4][1, Chap. 11]:

$$p = 0 \ , \ \frac{\partial p}{\partial t} = |\nabla p|^2 \ ;$$

the last condition can be replaced by

$$\frac{\partial p}{\partial t} = -V_n$$

where $V_n$ is the velocity of the free boundary. The Darcy law leads to some perhaps non-physical conclusions:

(i) As observed by John Hamilton and David Ross (in [1, Chap. 11]) the no-slip condition is not satisfied at the bottom point $A$ of the advancing free boundary; see Figure 3.5.

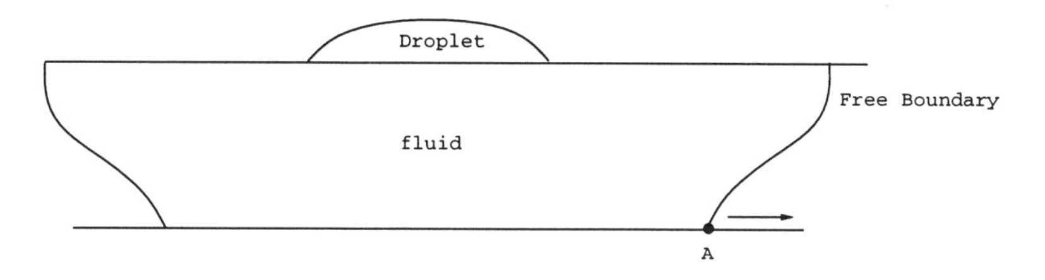


FIGURE 3.5.

(ii) According to L. Carson there does not seem to appear a sharp front (i.e., a crisp free boundary) as dictated by Darcy's law.

Because of the latter objection Carson has been using instead the concentration-gradient-driven flow. Accordingly,

$$\text{flux} \equiv c \ \vec{v} = -D\nabla c$$

where $c$ is the concentration of fluid and $D$ is a diffusion constant. Then the conservation of mass implies that

$$\frac{\partial c}{\partial t} = \nabla \cdot (D\nabla c) \ . \tag{3.5}$$

He assumes that the drop above the porous surface remains axially symmetric and that its contact area with the surface remains unchanged in time (i.e., no surface spreading) and he seeks an axially symmetric solution $c = c(r, z, t)$. The boundary conditions for this problem are as shown in Figure 3.6.

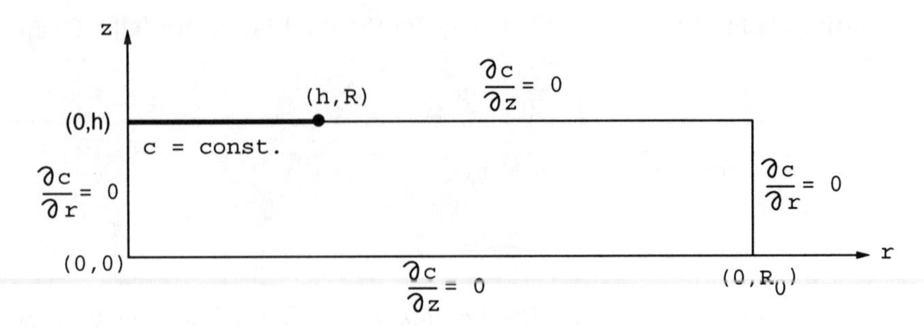


FIGURE 3.6.

The initial condition is

$$c(r, z, 0) = 0 \ .$$

He used scaled quantities:

distances in units of $h$ ,

dimensionless time $\tau = Dt/h^2$ , and

scaled concentration $\psi(x,t) = c(x,t)/f_{sat} \quad (x = (r, z))$

where $f_{sat}$ is the liquid volume fraction at saturation, i.e., porosity (In Figure 3.6 $\psi(r, h, t) = 1$ if $0 < r < R/h$). He used ANSYS to compute the solution for typical run parameters:

Time resolution of $\delta\tau \sim 10^{-2}$ ,

Spatial resolution of $\Delta r/h \sim \Delta z/h \sim 10^{-1}$ ,

Computational domain, $R_0/h \sim 10 - 20$ ,

Number of finite elements $\sim 1000 - 4000$ ,

with CPU time 5–10 hours on Convex C220. One of his results is shown in Figure 3.7; the notation $\langle\psi(r/h, z)\rangle$ indicates average with respect to the depth parameter $z$.

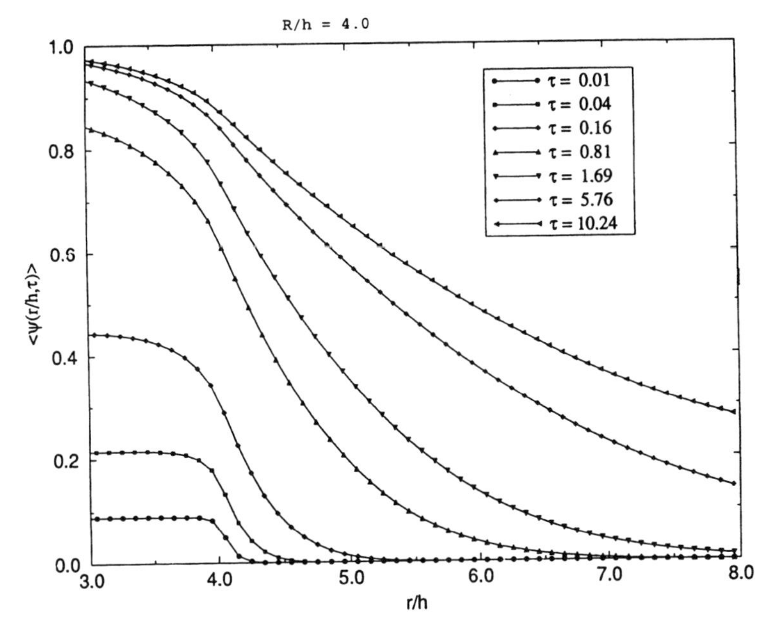


FIGURE 3.7.

The graphs for $\tau = 0.01$ and $\tau = 0.04$ indicate that the liquid flows primarily downward, saturating the cylindrical volume under the drop. As $\tau$ increases the direction of the flow becomes more outward: Thus, for $\tau \geq 1.69$ the liquid flows primarily outward from the saturated cylinder. These two modes of flow are illustrated in Figure 3.8, (a) and (b).

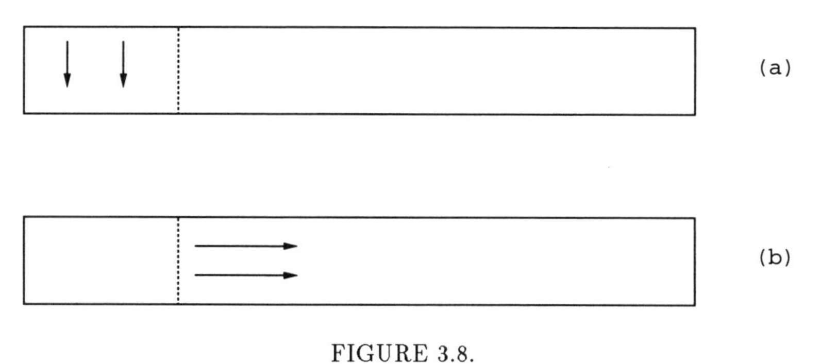


FIGURE 3.8.

## 3.4 Partially saturated porous media

In this section we introduce a third model which actually connects the two models discussed earlier, the pressure-driven flow (i.e., Darcy's law) and the concentration-gradient-driven flow (i.e., the diffusion model). The model is described in [4] and the solution is analyzed mathematically in [13].

We consider a nonstationary flow in a porous medium. We denote by $p$ the hydrostatic potential due to capillary suction, and by $u = c(p)$ the relative volumetric moisture content, i.e., the saturation. If $p$ exceeds a critical value, which we take to be 0, then the porous medium becomes saturated and $c(p)$ is equal to a constant, which we take to be 1. In the dry portion of the porous medium the pressure $p$ is constant and we take this constant to be $-1$. The function $c(p)$ is monotone increasing for $-1 < p < 0$ and (see [4, p. 488])

$$\frac{\partial c(p)}{\partial t} = \nabla \cdot (\kappa(p) \nabla p) \tag{3.6}$$

where $c(p) = 1,\ \kappa(p) = 1$ if $p > 0$; $c'(p) > 0, \kappa'(p) > 0$ if $p < 0$, $\quad c''(p) < 0$ if $-p_0 < p < 0$, $c''(p) > 0$ if $p < -p_0$ for some $p_0 > 1$, $c(p) \to 0$ and $\kappa(p) \to 0$ if $p \to -\infty$. Setting

$$K(p) = \int_0^p \kappa(q) dq, \quad \widetilde{c}(p) = c(K^{-1}(p))$$

we arrive at the equation

$$\frac{\partial}{\partial t} \widetilde{c}(\widetilde{p}) = \Delta \widetilde{p} \quad \text{for} \quad \widetilde{p} = K(p) \ .$$

Clearly $\widetilde{c}(s) = 1$ if $s > 0$, $\widetilde{c}'(s) > 0$ if $s < 0$. For many physically reasonable functions $(c(p),\ \kappa(p))$, there also holds:

$$\widetilde{c}''(s) \leq 0 \quad \text{if} \quad -1 < s < 0, \text{ and } \widetilde{c}(-1) \text{ is very small.}$$

In [13] it is assumed that $\widetilde{c}(-1) = 0$ and, for simplicity, that

$$\widetilde{c}(\widetilde{p}) = \begin{cases} 1 + \widetilde{p} & \text{if} \quad -1 < \widetilde{p} < 0 \\ 1 & \text{if} \quad \widetilde{p} \geq 0 \ ; \end{cases} \tag{3.7}$$

however, all the results obtained in that paper remain valid for $\widetilde{c}(p)$ satisfying:

$$\widetilde{c}'(p) > 0, \quad \widetilde{c}''(p) \leq 0 \quad \text{if} \quad -1 < p < 0 \ .$$

We note that in the regime $\{-1 < p < 0\}$ we have the diffusion model (used by L. Carson) and in the regime $\{p > 0\}$ we have Darcy's law.

We now briefly describe the results in [13] for the case (3.7), dropping for simplicity the tilda "$\sim$" over $\widetilde{p}, \widetilde{u}, \widetilde{c}$. We consider the 2-d case where the porous medium occupies the half space $\{y < 0\}$ and fluid penetrates into it from an interval $\{-a < x < a,\ y = 0\}$. The boundary conditions are:

$$\frac{\partial p(x, 0, t)}{\partial y} = -g(x, t)$$

where

$$g(x,t) = g(-x,t) \geq 0 ,$$

$$g(x,t) = 0 \quad \text{if} \quad |x| > a$$

and

$$p = -1 \quad \text{if} \quad x^2 + y^2 > A$$

where $A$ is a large constant.

We extend the definition of $c(p)$ to $p = -1$ by taking

$$c(p) = \min\{p, 0\} + H(p+1) \quad \text{if} \quad p \geq -1$$

where $H(s)$ is the Heaviside function, and take initial conditions

$$c(p(x,y,0)) = \gamma_0(x,y) = \gamma_0(-x,y)$$

where $\gamma_0(x,y) = -1$ if $x^2 + y^2$ is large. Note that $c(-1)$ is the set $[-1,0]$ and $c(p)$ is not defined (i.e., it is the empty set) for $p < -1$. Since the data are symmetric in $x$, we shall be looking for a solution which is also symmetric in $x$.

We note [14] that the equation

$$\frac{\partial}{\partial t} c(p) = \Delta p$$

across the wet/dry interface represents the physical equation of continuity (provided gravity is ignored) provided $c(p) \downarrow 0$ if $p > 0, p \downarrow 0$ and $c(p) = -1$ in the dry part of the porous medium.

For any $\rho > 0,\ T > 0$, set

$$\Omega_\rho = \{x^2 + y^2 < \rho^2,\ y < 0\} ,$$

$$\Gamma_{1\rho} = \{(x,0),\ -\rho < x < \rho\} ,$$

$$\Gamma_\rho = \{x^2 + y^2 = \rho^2,\ y < 0\}$$

and

$$\Omega_{R,T} = \Omega_\rho \times \{0 < t < T\},\ \Gamma_{1\rho,T} = \Gamma_{1\rho} \times \{0 < t < T\} ,$$
$$\Gamma_{\rho,T} = \Gamma_\rho \times \{0 < t < T\} .$$

*Definition.* A pair $(p,\gamma)$ is called a ***weak solution*** if

$$p \in L^2(0,T; H^1(\Omega_\rho)) \quad \text{for all} \quad \rho > 0,\ T > 0 ,$$

$$\gamma \in c(p) ,$$

$$p = -1 ,\ \gamma = -1 \quad \text{if} \quad x^2 + y^2 > \sigma(t)$$

where $\sigma(t)$ is a continuous monotone increasing function, and, for any $\rho > \sigma(T)$,

$$\iint_{\Omega_{\rho,T}} (\nabla p \cdot \nabla \zeta - \gamma \zeta_t) = \iint_{\Omega_\rho} \gamma_0 \zeta(\cdot, 0) + \iint_{\Gamma_{1\rho,T}} g\zeta \quad \forall\, \zeta \in V_{\rho,T}$$

where

$$V_{\rho,T} = \{\zeta \in H^1(\Omega_{\rho,T}),\ \zeta(\cdot, T) = 0\ ,\ \zeta|_{\Gamma_{\rho,T}} = 0\}\ .$$

Note that $c(p)$ is a graph, and $\gamma \in c(p)$ means that $\gamma(x,y,t) = c(p(x,y,t))$ if $p(x,y,t) > -1$,

$$-1 \leq \gamma(x,y,t) \leq 0 \quad \text{if} \quad p(x,y,t) = -1$$

and $p \geq -1$ almost everywhere (i.e., the graph $c(p)$ is empty if $p < -1$).

The following theorems are proved in [13].

**Theorem 3.1** *Under the above conditions there exists a unique weak solution.*

**Theorem 3.2** *If $\gamma_0 \in c(p_0)$ and $p_0, \gamma_0$ are continuously differentiable then the function $\max\{c(p),\ 0\}$ is continuous.*

**Theorem 3.3** *Suppose that*

$$\frac{\partial \gamma_0}{\partial x} \leq 0\ ,\ \frac{\partial \gamma_0}{\partial y} \geq 0\ ,\ \frac{\partial g}{\partial x} \leq 0 \quad \textit{if} \quad x > 0\ .$$

*Then there exist two continuous curves*

$$S_1 : r = \zeta_1(\theta, t)\ ,\ S_2 : r = \zeta_2(\theta, t) \qquad (0 \leq \theta \leq \pi\ ,\ t > 0)$$

*such that $0 < \zeta_1(\theta,t) < \zeta_2(\theta,t)$ and*

$$\begin{array}{ll} p > 0 & \textit{if} \ \ r < \zeta_1(\theta,t) \\ -1 < p < 0 & \textit{if} \ \ \zeta_1(\theta,t) < r < \zeta_2(\theta,t)\ , \end{array}$$

*and $p = -1$, $c(p) = 0$ if $r > \zeta_2(\theta,t)$.*

Thus the free boundary $S_1$ is the interface between the saturated region $\{p > 0\}$ and the partially saturated region $\{-1 < p < 0\}$, whereas the free boundary $S_2$ is the interface between the partially saturated region and the dry region; see Figure 3.9.

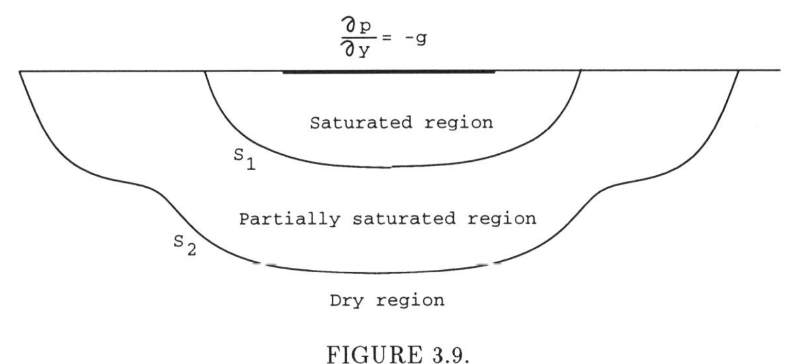


FIGURE 3.9.

## 3.5 Open problems

It would be interesting to examine more carefully the model of the partially saturated porous medium, i.e., the model introduced in Section 3.4. Some basic questions are:

*Problem (1).* Extend Theorems 3.1–3.3 to the case where the porous medium is an infinite slab $\{-\infty < x < \infty, -b < y < 0\}$.

Here the main difficulty arises when the free boundaries intersect the line $y = -b$.

*Problem (2).* Extend the results to the axially symmetric 3-d case.

*Problem (3).* Extend the results of Problem (1), (2) to rectangular domains with the same boundary conditions as in Section 3.3.

*Problem (4).* Compute the solution for the model of partially saturated porous medium for the same geometry and boundary conditions as in Section 3.3, and compare with the results of Carson, i.e., with Figure 3.7.

We finally note that although the wet/dry interface in the present model is a moving boundary, this boundary is not "crisp" since the moisture content of the fluid is very small near it. Thus this model does not seem to disagree with the observation (ii) in §3.3 made by L. Carson, although this model is different, in the undersaturated/dry regions, from his model (which excludes the dry region).

## 3.6 References

[1] A. Friedman, *Mathematics in Industrial Problems, Part 8*, IMA Volume 83, Springer–Verlag, New York (1996).

[2] R. Lukas, *Ueber das Zeitgesetz des Kapillaren Aufstiegs von Flüssigkeiten*, Kolloid Zeitschrift, 23 (1918), 15–22.

[3] E.W. Washburn, *The dynamics of capillary flow*, Phys. Review, 17 (2nd Series) (1921), 273–283.

[4] J. Bear, *Dynamics of Fluids in Porous Media*, American Elsevier, New York (1972).

[5] H.C. Brinkman, *A calculation of the viscous force exerted by a flowing fluid on a dense swarm of particles*, Appl. Sci. Res., A1 (1947), 27–34.

[6] G. Allaire, *Homogenization of the Navier–Stokes equations in open sets perporated with tiny holes*, I: Abstract framework, a volume distribution of holes; II: Non-critical sizes of the holes for a volume distribution and a surface distribution of holes, Archive Rat. Mech. Anals., 113 (1991), 209–259; 261–298.

[7] J.B. Keller, *Darcy's law for flow in porous media and the two-phase method*, Lecture Notes in Pure and Appl., Math., 54, Dekker, New York (1980).

[8] J.L. Lions, *Some Methods in the Mathematical Analysis of Systems and Their Control*, Beijing, Gordon and Breach, New York (1981).

[9] E. Sanchez–Palencia, *Non Homogeneous Media and Vibration Theory*, Lecture Notes in Physics, 127, Springer–Verlag, Heidelberg (1980).

[10] L. Tartar, *Convergence of the homogenized process*, Appendix of [7].

[11] G. Allaire, *Homogenization of the Stokes flow in a connected porous medium*, Asymptotic Analysis, 2 (1989), 203–222.

[12] G. Allaire, *Homogenization of the Navier–Stokes equations with a slip boundary condition*, Comm. Pure Appl. Math., 44 (1991), 605–641.

[13] X. Chen, A. Friedman, and T. Kimura, *Nonstationary filtration in partially satured porous media*, European J. Appl. Math., 5 (1994), 405–429.

[14] E. DiBenedetto and A. Friedman, *Periodic behavior for the evolutionary dam problem and related free boundary problems*, Comm. P.D.E., 11 (1986), 1297–1377.

# 4

# Information theory and sensor management

Many industrial and military surveillance problems entail the use of sensors that must be managed in various ways. Typical applications include using a camera to direct a robotic arm to grasp objects of a particular type from a conveyor belt and automatic management of radars and imaging sensors on military vehicles. For example, in the robot problem many decisions must be made automatically about how to direct the camera effectively. It can be directed at different points on the belt and it can have multiple modes such as different magnification settings and different spectral bands to select from. Before an object can be grasped there are several steps that must be accomplished by the robot. First, objects must be detected on the belt. Second, objects must be classified as to type. Finally, once an object of the correct type has been selected, it must be tracked and its orientation must be determined so that the manipulator can be controlled. Optimal performances for each of these tasks leads to different requirements for use of the sensor. Often, different types of tasks will be performed simultaneously. This is the case when some objects are being detected and classified while another object is grasped.

Although some underlying principles are beginning to emerge and rigorous results have been obtained for few simplified situations, there is no general basis for solving automation problems of this type. On November 10, 1995 Keith Kastella from LORAL Corporation described a heuristic approach to sensor management based on maximizing expected information gain. A key feature of this approach is that the sensor manager is incorporated into some kind of filtering process. The filter includes a prediction model that enables the information gain from alternative sensor modes and dwell points to be estimated. The expected optimal sensor mode and dwell points can then be selected. This approach has been applied in [1][2] to a multisensor/multitarget radar tracking problem and to a detection/classification problem.

## 4.1 The discrimination function

One of the problems facing designers of sensor management is that there is little consensus what mathematical quantities should be optimized; see Kastella [2]. Much of the work in this area has focused on problems where the data are primarily kinematic, with little attention on the detection/classification aspects of the problem. In these applications the Kalman filter is used to predict how state estimates will evolve in time based on the estimated target dynamics and signal-to-noise ratio; see Section 4.2 below. Schmaedeke and Kastella [1] [2] developed a method for managing agile sensors to optimize detection and classification based on discrimination gain. The method is a heuristic based on generating the *least likely* densities for the probability density function (PDF) of the target. This heuristic resembles the strategy used in 20 questions game, where one selects questions that maximize the expected information gain. Questions whose answers are quite certain have little value while questions whose answers are quite uncertain are more valuable.

In this section we shall define the discrimination function that measures the relative likelihood of different probability densities, and then explain the motivation for choosing the maximal discrimination gain for the sensor.

To define the discrimination function consider the problem of classifying objects that exist in $S$ mutually exclusive discrete classes. Let $P(s)$ and $Q(s)$ be two probabilities for the objects of $S$, labeled by $s = 1, 2, \ldots, S$. Then the *discrimination* of $P$ with respect to $Q$ is defined by

$$D(P,Q) = \sum_{s=1}^{S} P(s) \ln \frac{P(s)}{Q(s)} \; ; \tag{4.1}$$

this is also called the Kullback–Leibler information, or *cross-entropy* [3]. It can be shown that $D(P,Q)$ is nonnegative and is 0 if and only if $P = Q$. $D(P,Q)$ is an average of the log-likelihood $\ln[P(s)/Q(s)]$, and has been used in hypothesis testing problems [4].

Let us illustrate the role of discrimination-based sensor management in a problem of target detection/classification for a single detection cell. The true state of the cell, denoted by $\tilde{s}$, is one of the $S$ states, and

$\tilde{s} = 1$ denotes no target,

$\tilde{s} = 2, \ldots, s$ denotes one of possible target types.

There is at most one target in the cell and, based on prior information, the probability that a cell is in state $s$ is given by the prior distribution $Q(s)$. Denote by $P(s)$ the estimated probability that the cell is in state $s$, based on some measurements. If $P = Q$ then we have obtained no new information, and in this case $D(P,Q) = 0$. If $Q$ is uniformly distributed then perfect information (i.e., $P(s) = 1$ if $s = \tilde{s}$ and $P(s) = 0$ otherwise), which is not very likely given the prior $Q$, gives the maximum value of the discrimination of $P$ with respect to $Q$.

Consider next a situation where several targets or objects have to be selected. Suppose $n$ objects are drawn at random from an infinite set of objects, labeled by a finite set of indices $s = 1, \ldots, S$. On each draw the probability of selecting an object labeled $s$ is given by the prior probability $Q(s) \equiv q_s$. Then the probability of drawing $n_s$ objects labeled $s$ is given by the multinomial distribution

$$P(n_1, \ldots, n_S) = \frac{n!}{n_1! \ldots n_S!}$$

where $\sum_{s=1}^{S} n_s = n$. Defining $P(s) = n_s/n$ for each $s$ and using Stirling's formula, one can show that [4]

$$P(n_1, \ldots, n_S) = e^{nD(P,Q)+o(n)} \quad \text{as} \quad n \to \infty \, . \tag{4.2}$$

We see that if the distribution $P$ is very unlikely then $D(P, Q)$ is large, and it has large information content relative to the a priori information. Thus $D(P, Q)$ seems to provide a plausible measure of information contained in $P$: If the increase in $D(P, Q)$ is large we gain more information on the unlikelihood of the event $(n_1, \ldots, n_S)$, whereas if the change in $D(P, Q)$ is small we gain little information over the prior information embodied in $Q$.

In Section 4.3 we shall give an example of multitarget/multisensor track optimization based on [2], and in Section 4.4 we shall give an example of detection/classification on a grid. Although these are two different problems, the discrimination approach will be used in both cases.

## 4.2 The Kalman filter

The results of Section 4.3 use the Kalman filter which predicts target uncertainty for sensor/target pairing. We shall therefore describe in this section the Kalman filter; for more details we refer to [5] or [6, Chap. 2]. A standard reference in the engineering literature is [7].

A process $x_k$ evolves in discrete time steps $k = 0, 1, \ldots$, according to

$$x_k = A_k \; x_{k-1} + N_k \qquad \text{(the state equation)}, \tag{4.3}$$

where $A_k$, the state transition matrices, are known and $N_k$, the driving noises, are random variables. It is assumed that the $N_k$ are independent Gaussian variables with mean zero and known covariances

$$Q_k = E(N_k N_k^T) \, .$$

At each time $k$ we can measure $x_k$ only indirectly and with some error $\varepsilon_k$; more precisely, we measure a quantity $z_k$ given by

$$z_k = H_k x_k + \varepsilon_k \, . \tag{4.4}$$

$H_k$, the output sensitivity matrix, is given. The $\varepsilon_k$ are assumed to be independent Gaussian variables with mean zero and known covariances

$$R_k = E(\varepsilon_k \varepsilon_k^T) \, ,$$

which are nonsingular. We further assume that the $\{N_k\}$ and $\{\varepsilon_k\}$ are independent.

The problem is how to optimally estimate the state $x_k$ or, in other words, how to optimally track the target $x_k$. This is achieved by the Kalman filter which will be described below.

We first define two processes:

$\widehat{x}_{j|j}$ = the optimal estimate we wish to construct, at time $j$ ,

$\widehat{x}_{k|k-1}$ = the forecast state which will be automatically chosen once we have determined $\widehat{x}_{k-1|k-1}$ .

$\widehat{x}_{j|j}$ is naturally defined as the conditional probability of $x_j$ with respect to the variables $z_0, z_1, \ldots, z_j$. $\widehat{x}_{k|k-1}$ is naturally defined by

$$\widehat{x}_{k|k-1} = A_k \widehat{x}_{k-1|k-1} \quad \text{(the prediction equation).}$$

We introduce two matrices:

$$P_{j|j} = E[(x_j - \widehat{x}_{j|j})(x_j - \widehat{x}_{j|j})^T] \quad \text{(the update covariance),}$$

$$P_{j|j-1} = E[(x_j - \widehat{x}_{j|j-1})(x_j - \widehat{x}_{j|j-1})^T] \quad \text{(the prediction covariance)}$$

One can easily verify that

$$P_{k|k-1} = A_k P_{k-1|k-1} A_k^T + Q_k \, . \tag{4.5}$$

One can also compute $\widehat{x}_{k|k}$ by

$$\begin{aligned} \widehat{x}_{k|k} &= \widehat{x}_{k|k-1} + K_k(z_k - H_k \widehat{x}_{k|k-1}) \\ &= \widehat{x}_{k|k-1} + K_k(z_k - H_k A_k \widehat{x}_{k-1|k-1}) \end{aligned} \tag{4.6}$$

where the matrix $K_k$ satisfies:

$$P_{k|k} = P_{k|k-1} - K_k(N_k A_k P_{k-1|k-1} A_k^T + H_k Q_k) \, , \tag{4.7}$$

and

$$K_k = (A_k P_{k-1|k-1} A_k^T + Q_k) H_k^T (H_k (A_k P_{k-1|k-1} A_k^T) H_k^T + R_k)^{-1} \, . \tag{4.8}$$

$K$ is called the *Kalman gain* or the *Kalman filter*, and (4.6) is called the *Kalman filter equation*, or the state measurement update equation. For this linear case, $\widehat{x}_{k|k}$ is the minimum mean-square error estimate.

Observe that the Kalman filter can be constructed successively from the data $A, H, Q, R$ as follows: Given $P_{k-1|k-1}$ and $K_k$ we first compute $P_{k|k-1}$ from (4.5), and then $P_{k|k}$ from (4.7) and $K_{k+1}$ from (4.8).

From the above formulas we can also derive an expression for $K_k$ in terms of $P_{k|k-1}$ (see [6, p. 14]),

$$K_k = P_{k|k-1}H_k^T(H_kP_{k|k-1}H_k^T + R_k)^{-1} . \tag{4.9}$$

We can use (4.5) to simplify (4.7):

$$P_{k|k} = P_{k|k-1} - K_kH_kP_{k|k-1} = (I - K_kH_k)P_{k|k-1} .$$

Hence

$$(P_{k|k})^{-1} = (P_{k|k-1})^{-1}(I - K_kH_k)^{-1} . \tag{4.10}$$

From (4.9) we get

$$K_kH_k = (I + H_k^{-1}R_k(H_k^T)^{-1}(P_{k|k-1})^{-1}) \equiv (I + T)^{-1}$$

so that

$$(I - K_kH_k)^{-1} = (I - (I+T)^{-1})^{-1} = I + T^{-1} .$$

Then, by (4.10),

$$(P_{k|k})^{-1} = (P_{k|k-1})^{-1}(I + P_{k|k-1}H_k^TR_k^{-1}H_k) ,$$

or

$$(P_{k|k})^{-1} = (P_{k|k-1})^{-1} + H_k^TR_k^{-1}H_k . \tag{4.11}$$

This formula will be used in the next section.

## 4.3 Multitarget/multisensor track optimization

We assume that there are $S$ sensors with measurements which entail Gaussian errors having covariances $R_s$ $(s = 1, \ldots, S)$, and $T$ targets with Gaussian noise having covariances $Q_t$, $t = 1, \ldots, T$.

We can apply the Kalman filter with $A_k$, $H_k$ the identity matrices. Denote by $\Sigma(\text{old})$ the covariance $P_{k-1|k-1}$ and by $\Sigma(\text{new})$ the covariance $P_{k|k}$. Then, by (4.5),

$$\Sigma(\text{new}) = \Sigma(\text{old}) + Q_t \tag{4.12}$$

if target $t$ is not observed, and, by (4.11),

$$(\Sigma(\text{new}))^{-1} = (\Sigma(\text{old}) + Q_t)^{-1} + R_{i_1}^{-1} + \cdots + R_{i_n}^{-1} \tag{4.13}$$

if target $t$ is observed by the sensors in the set $u = \{i_1, \ldots, i_n\}$.

Suppose $f(x)$ $(x \in X)$ is PDF which represents the relative information content with respect to non-informative PDF $\lambda(x)$. Then

$$I(f,\lambda) \equiv \int_X f(x) \ln \frac{f(x)}{\lambda(x)}\, dx$$

is called the *informative content* of $f$ (Tarantola [8, p. 28]). (This concept coincides with that of the discrimination function if $X$ is a discrete space.) If $f_1, f_2$ are two PDF's representing different relative information, then the *gain* in the information of $f_2$ over $f_1$ is defined by

$$G = I(f_2,\lambda) - I(f_1,\lambda) .$$

If $f_1, f_2$ and $\lambda$ are Gaussian with means $m_1, m_2, m_0$ and covariances $C_1$, $C_2, C_0$ respectively, then [7, p. 153]

$$\begin{aligned} G = \frac{1}{2} \ln \frac{\det(C_1)}{\det(C_2)} &+ \frac{1}{2}(m_2 - m_0)^T C_0^{-1}(m_2 - m_0) \\ &- \frac{1}{2}(m_1 - m_0)^T C_0^{-1}(m_1 - m_0) + \frac{1}{2} \operatorname{Trace}[(C_2 - C_1)C_0^{-1}] . \end{aligned} \tag{4.14}$$

We can use this formula to compute the gain in information when target $t$ is observed by a set of sensors $u = (i_1, \ldots, i_n)$; the covariant matrices $C_1, C_2$ are given respectively by (4.12) and (4.13).

Let $S^\#$ denote the set of all subsets $u$ of sensors. Let $J(s)$ denote the set of all subsets of $S$ which contain $s$. Based on the heuristic strategy of §4.1 we wish to

$$\text{maximize } I = \sum_{u \in S^\#} \sum_{t=1}^{T} G_{ut} x_{ut} \tag{4.15}$$

subject to

$$\sum_{u \in S^\#} x_{ut} \le 1 \quad \text{for} \quad t = 1, \ldots, T , \tag{4.16}$$

$$x_{ut} \ge 0 , \tag{4.17}$$

$$\sum_{u \in J(s)} x_{ut} \le \tau(s) , \quad s = 1, \ldots, S . \tag{4.18}$$

Here $G_{ut}$ is the gain when a set $u$ of sensors is directed to observe target $t$, and is computed by (4.14) and (4.12), (4.13), and $\tau(s)$ is the capacity of sensor $s$. Since the objective function in (4.15) is linear, the maximum is achieved on the boundary, $x_{ut} = 0$ or 1. When $x_{ut} = 1$, then sensor set $u$ is assigned to target $t$. Thus (4.18) mean that the total commitment for all sensor groups which contain sensor $s$ cannot exceed the capacity of $s$.

The optimization approach requires the solution of linear programming problem at each compute step and therefore may be too compute intensive.

One alternative is a greedy approach in which one first seeks the maximum information gain element $G_{ut}$ in the gain matrix. Then the remaining elements in the maximizing target $t$'s column are set to zero (so no sensor is assigned more than once to the same target on the same step), and the sensor capacities $\tau(s)$ for the sensors $s$ in the sensor set $u$ are decremented by 1. The maximum is sought from the remaining targets with the modified capacities. This process is repeated until the sensor capacity has been used.

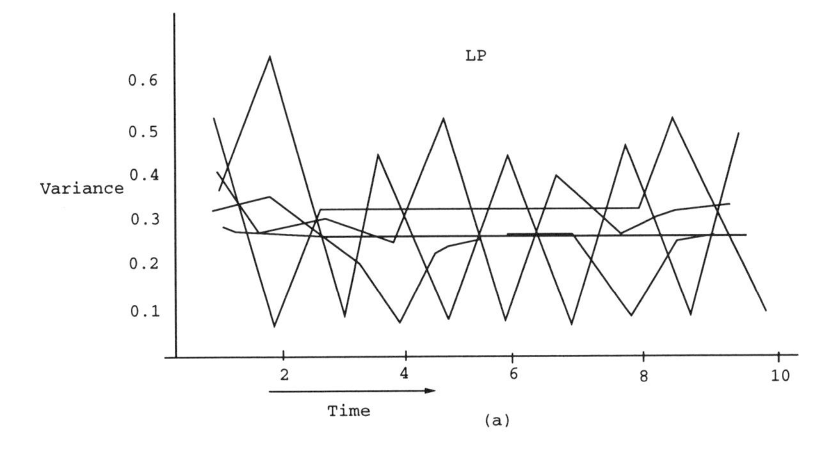


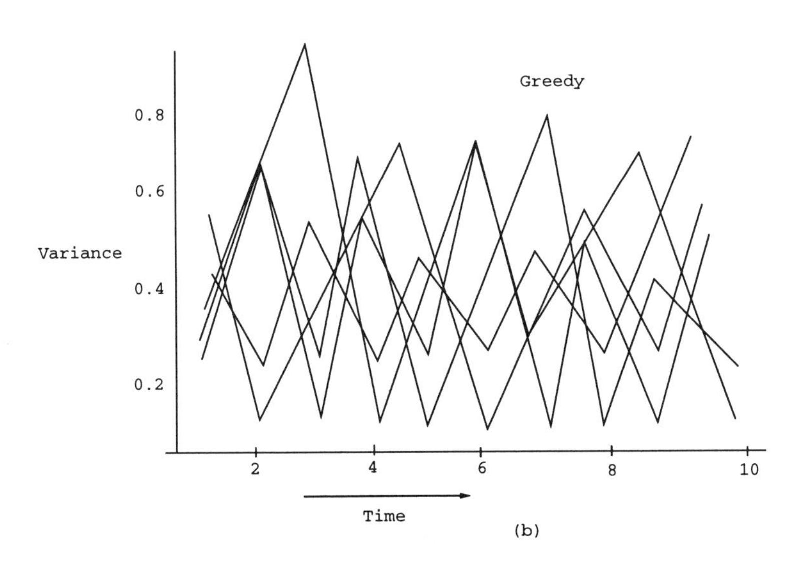


FIGURE 4.1.

Figure 4.1 shows an example taken from [2] for the case of three sensors and five targets The sensor variances are 0.35, 0.1, 0.6 and the process variances are 0.2, 0.25, 0.3, 0.35, 0.4. As time evolves, the $LP$ method shows smaller variances in tracking the five targets than the variances obtained by the greedy method.

## 4.4 Detection/classification on a grid

Consider a collection of cells indexed by $c = 1, 2, \ldots, C$. In each cell there is at most one target. There are $S - 1$ target types. Thus we can think of each cell as being in one of $S$ states $s = 1, 2, \ldots, S$ where $s = 1$ means no target in the cell. The system is observed by a sensor which can be directed toward any single cell without error. At each time-step only one cell is sampled, producing a measurement $z$. The observation outcomes are i.i.d. The conditional probabilities $P_c[z|s]$ to obtain $z$ given that cell $c$ is in state $s$ are known. Also, a priori probabilities $P_c[s]$ for $c$ to be in state $s$ are known.

The detection/classification problem is to determine in which state the cell is, based on a set of observations. We shall follow here the approach of Kastella [1].

Denote the outcome of the $k$-th observation of cell $c$ by $z_k$ and set $Z_K = (z_1, z_2, \ldots, z_K)$. Then after $K$ observations the posterior probability $P_c[s|Z_K]$ that $z$ is in state $s$ conditioned on the observations can be computed by Bayes formula:

$$P_c[s|Z_K] = P_c[s] \frac{\prod_{k=1}^{K} P_c[z_k|s]}{\sum_{t=1}^{S} P_c[t] \prod_{k=1}^{K} P_c[z_k|t]}$$

or, by the recursive formula,

$$P_c[s|Z_k] = \frac{P_c[z_K|s] P_c[s|Z_{K-1}]}{\sum_{t=1}^{S} P_c[z_K|t] P_c[t|Z_{K-1}]} . \tag{4.19}$$

For each cell $c$ the minimum-probability-of-error detector is obtained by choosing $s$ to maximize $P_c[s|Z_K]$ [4]. Thus it remains to decide how to optimally choose the next cell.

After $K$ observations, the discrimination of the estimated probability distribution with respect to the prior $P_c[s]$ is

$$D_c[Z_K] = \sum_{s=1}^{S} P_c[s|Z_K] \ln \frac{P_c[s|Z_K]}{P_c[s]} . \tag{4.20}$$

The probability density for the $K$-th observation can be computed by

$$P_c[z_K|Z_{K-1}] = \sum_{s=1}^{S} P_c[z_K|s]P_c[s|Z_{K-1}] \tag{4.21}$$

since the states are mutually exclusive. Therefore the expected discrimination after one additional observation is

$$E_c[D|Z_{K-1}] = \sum_{z_K} D_c[Z_K]P_c[z_K|Z_{K-1}] \,. \tag{4.22}$$

The expected discrimination gain for cell $c$ is then

$$\Delta D_c[Z_K] = E_c[D|Z_{K-1}] - D_c[Z_K] \,. \tag{4.23}$$

The discrimination-based sensor management strategy, using the heuristic of §4.1, is to select the cell $c$ that maximizes $\Delta D_c[Z_K]$.

The above scheme can be used, for example, on a grid of cells for detection only. In this situation each cell can be in one of two states, i.e., either the cell contains a target or it does not. We are interested in detecting the correct state of the cell, whose probability will be denoted by $P_d$. False alarm means that we determined that there is a target in the cell while the truth is that there is no target in the cell; the probability of false alarm will be denoted by $P_{fa}$.

Given prior $P_d, P_{fa}$ the above scheme was computed in [1] in the case of 100 cells with initial $P_d = 0.8$ , $P_{fa} = 0.5$; after 50 trials with 10 samples/cells the result was $\widetilde{P}_d = 0.9$, $P_{fa} = 0.001$. Direct search yielded the numbers $\widetilde{P}_d = 0.1$ , $\widetilde{P}_{fa} = 0.01$. Thus the discriminant-based search gives a substantial improvement in accuracy.

Ongoing work by Keith Kastella, Troy Jenison, Stan Musick and Wayne Schmaedeke deals with multitarget/multisensor detection and classification. They considered situations where some sensors are better equipped for detection while others are better equipped for classification. Numerical examples then show that at small times the detection sensors are used more frequently than the classification sensors.

Some basic questions for future research are:

(i) To define the concept of optimal detection/classification strategy, a concept that might include simultaneous employment of sensors.

(ii) How good is the heuristic method based on the discrimination function?

(iii) How to extend the results to the case of an unknown number of targets?

## 4.5 References

[1] W. Schmaedeke, *Information-based Sensor Management*, SPIE Proceedings Vol. 1955, Signal Processing, Sensor Fusion, and Target Recognition II, April, 1993. W. Schmaedeke and K. Kastella, *Information Theory and Sensor Management*, Basic Research Group of the Joint Directors of Laboratories 1993 Symposium on Command and Control Research, June 28–July 1, 1993.

[2] K. Kastella, *Discrimination gain to optimize detection and classification*, IMA Preprint Series # 1288, February 1995. To appear, IEEE Trans. Sys, Man, Cyber.

[3] J.N. Kapur and H.K. Kesavan, *Entropy Optimization Principles with Applications*, Academic Press, Boston (1992).

[4] R.E. Blahut, *Principles and Practice of Information Theory*, Addison–Wesley, Reading, MA (1987).

[5] R.S. Liptser and A.N. Shiryayev, *Statistics of Random Processes*, Volumes 1,2, Springer–Verlag, New York (1977), (1978).

[6] A. Friedman, *Mathematics in Industrial Problems, Part 6*, IMA Volume 57, Springer–Verlag, New York (1993).

[7] A.H. Jazwinski, *Stochastic Processes and Filtering Theory*, Academic Press, New York, NY (1970).

[8] A. Tarantola, *Inverse Problem Theory*, Elsevier, Amsterdam (1987).

# 5

# Problems and applications in density-functional theory

A principle goal of modern chemical research is the rational design of chemicals and materials through the manipulation of molecular properties at the atomic level. The role of computational chemistry in this endeavor is to provide a method of obtaining atomic-level properties through calculation rather than more expensive and time-consuming experiments. Computational chemistry is still in a relative infancy, and the properties that one can calculate with useful accuracy are still limited. They include things such as molecular geometries, relative conformational energies, binding energies, charge distributions, and excitation energies. On November 17, 1995 Bill Schneider from Ford Motor Company described three applications of computational chemistry that he has worked on: (i) catalysts for selective reduction of $NO_x$ for new catalytic converter applications (ii) atmospheric degradation of hydrofluorocarbons (HFCs), which are replacing chlorofluorocarbons (CFCs) in automotive air conditioners; and (iii) the photochemistry of ultraviolet absorbers in automotive paints.

These applications present a variety of unique challenges to computational chemistry, in particular to the relative new methods based on density functional theory. Density functional theory holds the promise of making tractable much larger chemical systems than can be treated using traditional ab initio (first principles) techniques, but the theory is still in a state of development. This chapter will provide background on the traditional ab initio techniques, then describe density functional theory, and finally comment on the applicability of density functional theory to the aforementioned applications. More details of density functional theory can be found in the fairly recent books [1–3].

## 5.1 The Schrödinger equation

Suppose a molecule is composed of two atoms, $A$ and $B$. What are the stationary states of the molecule as a function of the separation $A$ and $B$? This question is one of the most fundamental problems that computational chemists want to answer by solving the Schrödinger equation. If a particle with mass $m$ and position $\mathbf{r}$ is moving under the influence of a potential

$V(\mathbf{r})$, the Schrödinger equation reads:

$$\frac{\hbar}{i}\frac{\partial\psi(\mathbf{r},t)}{\partial t} = \mathbf{H}\psi(\mathbf{r},t)$$

where $\mathbf{H}$ is the Hamiltonian operator

$$\mathbf{H} = -\frac{\hbar^2}{2m}\nabla^2 + V(\mathbf{r}) \, .$$

Here $\hbar$ is Planck's constant divided by $2\pi$, and is approximately $1.05\times 10^{-27}$ erg. sec. The solution $\psi(\mathbf{r},t)$ is generally complex-valued and $|\psi|^2$ is the probability distribution for the position of the particle. One is particularly interested in time-harmonic solutions (i.e., solutions of the form $e^{i\lambda t}\psi(\mathbf{r})$) and, in fact, in the eigenfunctions of $\mathbf{H}$. The eigenfunction of $\mathbf{H}$ with the smallest eigenvalue is called the *ground state* of the molecule.

Consider now the case where we have an $n$-particle system, i.e., a system defined by $n$ independent particles with coordinates $q_1,\ldots,q_n$. In this case the Hamiltonian is

$$\mathbf{H} = -\hbar^2\sum_{i=1}^{n}\frac{\nabla_i^2}{2m_i} + V(q_1,\ldots,q_n) \tag{5.1}$$

where $m_i$ is the mass of $q_i$; the first term represents the kinetic energy and $V$ represents the potential energy.

The eigenfunctions $\Psi_j = \Psi_j(q_1,\ldots,q_n)$ (also called eigenvectors or wave-functions) satisfy the equations:

$$\mathbf{H}\Psi_j = E_j\Psi_j \, . \tag{5.2}$$

Here $E_j$ denote the discrete sequence of energy levels of the eigenstates, the $\Psi_j$ are chosen to form a complete orthonormal sequence in the Hilbert space $L^2$, and $|\Psi_j|^2 dq$ is the probability of finding the system in volume $dq$ when in energy state $E_j$. For fermions (such as electrons) the $\Psi_j$ are antisymmetric to particle permutation.

Consider the simple example of a hydrogen atom. Because the mass of the nucleus is much larger than the mass of the electron, we can take the nucleus as fixed at the center of mass, and the Hamiltonian becomes

$$H = -\frac{1}{2}\nabla^2 - \frac{1}{r} \, ,$$

after scaling. Hence the eigenvectors are determined by

$$\left[-\frac{1}{2}\nabla^2 - \frac{1}{r}\right]\Psi_n(r) = E_n\Psi_n(\mathbf{r}) \, .$$

The eigenfunctions ("orbitals") seperate into products

$$\Psi_{n\ell m_\ell}(\mathbf{r},\theta,\phi) = R_{n\ell}(\mathbf{r})Y_{\ell m_\ell}(\theta,\phi)$$

of Laguerre polynomials and spherical harmonics. They are labeled by three "quantum numbers": principle, $n = 1, 2, \ldots$; azimuthal, $\ell = 0, 1, n-1$; and magnetic, $m_\ell = -\ell, \ldots, \ell$. A relativistic treatment introduces a fourth quantum number—electron spin, $m_s = \pm 1/2$—that is introduced ad hoc in non-relavistic treatments. Figures 5.1a and 5.1b depict the radial parts of the eigenfunctions for the states $n = 1$, $\ell = 0, m_\ell = 0$ and $n = 2, \ell = 0, m_\ell = 0$, respectively.

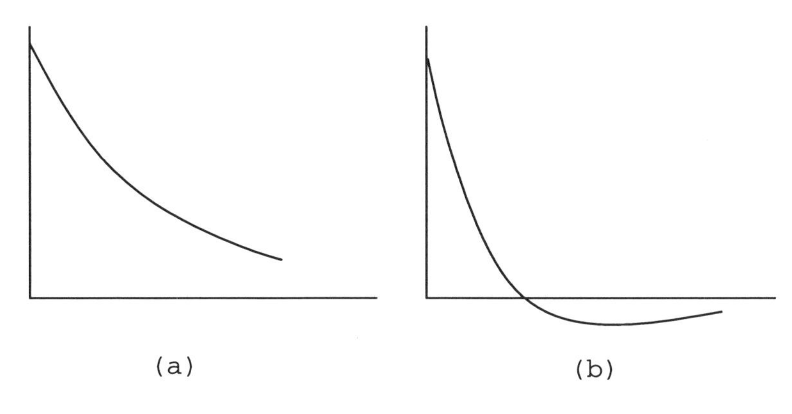


FIGURE 5.1.

Consider next a molecule composed of $N$ atoms with nuclear charges $Z_a$ and $n$ electrons (the electron charge is scaled to -1). In the Born-Oppenheimer approximation, the motions of the relatively massive nuclei are assumed to be decoupled from that of the much lighter electrons; that is, the nuclei are treated as moving in a potential field generated by the electronic wavefunction. After rescaling the masses, the Schrödinger equation for the molecule becomes:

$$\mathbf{H}\Psi_j(\mathbf{r}_1, \ldots, \mathbf{r}_n; \mathbf{R}_1, \ldots, \mathbf{R}_N) = E_j \Psi_j(\mathbf{r}_1, \ldots, \mathbf{r}_n; \mathbf{R}_1, \ldots, \mathbf{R}_N) \ ,$$

$$\mathbf{H} = -\frac{1}{2}\sum_{i=1}^{n} \nabla_i^2 - \sum_{a=1}^{N}\sum_{i=1}^{n} \frac{Z_a}{r_{ia}} + \sum_{1 \leq i < j \leq n} \frac{1}{r_{ij}} + \sum_{1 \leq a < b \leq N} \frac{Z_a Z_b}{R_{ab}} \tag{5.3}$$

where $r_{ia}$ is the distance of the $i$-th electron from the $a$-th nucleus, i.e. $r_{ia} = |\mathbf{r}_i - \mathbf{R}_a|$, $r_{ij}$ is the distance from the $i$-th electron to the $j$-th electron, i.e., $r_{ij} = |\mathbf{r}_i - \mathbf{r}_j|$, and $R_{ab}$ is similarly the distance from the $a$-th nucleus to the $b$-th nucleus. Note in the Born-Oppenheimer approximation the wavefunction and energy depend parametrically on the nuclear positions. Also observe that the electronic motion is strongly coupled ("correlated") through the Coulomb potential term (third term on the right). The problem (5.3) is $3n$-dimensional; in general it has no closed form solutions for $n > 1$, and it is difficult to calculate accurate solutions unless $n$ is quite small ($n \leq 10$). Fortunately, for most problems of chemical interest one is only interested in relative energies as, say, the $\mathbf{R}_j$ are varied, and absolute accuracy is not essential.

Chemists have devised a variety of techniques for solving the molecular Schrödinger equation in the Born-Oppenheimer approximation. The most common is the Hartree-Fock, or independent electron, approximation. The Hartree-Fock method and improvements to it, all of which attempt to solve (5.3) directly, are called ab initio (i.e., from first principles). Highly accurate (for chemical applications) ab initio solutions are tractable for systems up to $n \approx 50$, and less accurate Hartree-Fock solutions for systems up to $n = 500$. Further approximations to the Hartree-Fock approach, based on parameterizations against experimental data, are called semi-empirical, and can treat systems an order of magnitude larger. Density functional theory provides a completely independent approach to the problem posed in (5.3), but to understand this theory, it is useful first to examine the basis of the Hartree-Fock approach.

## 5.2 The Hartree–Fock method

The Hartree–Fock method attempts to reduce the $(3n)$-dimensionality of the Schrödinger equation by seeking wavefunctions $\Psi(\mathbf{r}_1, \ldots, \mathbf{r}_n)$ of the form

$$\Psi(\mathbf{r}_1, \ldots, \mathbf{r}_n) = \frac{1}{\sqrt{n!}} \begin{vmatrix} \psi_1(\mathbf{r}_1) & \ldots & \psi_n(\mathbf{r}_1) \\ \vdots & \ddots & \vdots \\ \psi_1(\mathbf{r}_n) & \ldots & \psi_n(\mathbf{r}_n) \end{vmatrix} . \tag{5.4}$$

The determinant on the right-hand side is called the "Slater determinant." The $\psi_i$ are one electron wavefunctions, or orbitals. The determinant form is used to ensure the antisymmetry of the wavefunction required of fermions. While factorization of $\Psi$ into one-electron orbitals is obviously a severe approximation, their introduction leads to simplification of (5.3) into an effective one-electron Hamiltonian that is readily solved.

To find the ground state we need to solve the variational problem:

$$\begin{aligned} &\text{minimize} \quad E_0 = \int \Psi^* \mathbf{H} \Psi \\ &\text{subject to} \quad \int \psi_i^* \psi_j = \delta_{ij} \end{aligned} \tag{5.5}$$

where "*" stands for complex conjugate. This leads to the Hartree–Fock (one electron) equation

$$f(\mathbf{r}_1)\psi_i(\mathbf{r}_1) = \varepsilon_i \psi_i(\mathbf{r}_1) \tag{5.6}$$

where, in the case of $\mathbf{H}$ as in (5.3),

$$f(\mathbf{r}_1) = -\frac{1}{2} \nabla_1^2 - \sum_a \frac{Z_a}{r_{1a}} + \sum_j [J_j(\mathbf{r}_1) - K_j(\mathbf{r}_1)] \; ; \tag{5.7}$$

here

$$J_j(\mathbf{r}_1)\psi_i(\mathbf{r}_1) = \psi_i(\mathbf{r}_1) \int |\psi_j(\mathbf{r}_2)|^2 \frac{1}{r_{12}}\, d\mathbf{r}_2 \ ,$$
$$J_{ij} = \int \psi_i^*(\mathbf{r}_1) J_j(\mathbf{r}_1)\psi_i(\mathbf{r}_1) d\mathbf{r}_1 \tag{5.8}$$

are the Coulomb operators and integrals representing repulsion between electron distributions $|\psi_i|^2$ and $|\psi_j|^2$, and

$$K_j(\mathbf{r}_1)\psi_i(\mathbf{r}_1) = \psi_j(\mathbf{r}_1) \int \psi_j^*(\mathbf{r}_2)\psi_i(\mathbf{r}_2) \frac{1}{r_{12}}\, d\mathbf{r}_2 \ ,$$
$$K_{ij} = \int \psi_i^*(\mathbf{r}_1) K_j(\mathbf{r}_1)\psi_i(\mathbf{r}_1) d\mathbf{r}_1 \tag{5.9}$$

are the *exchange* operators and integrals that arise from the anti-symmetry condition. These operators have no classical analog. Finally, the $\varepsilon_i$ in (5.6) are the Lagrange multipliers which arise from the constraints in (5.5), and

$$E_{HF} = \sum_i \varepsilon_i - \sum_{i>j} (J_{ij} - K_{ij}) \tag{5.10}$$

is the Hartree-Fock energy.

With today's computing power chemists can calculate solutions to the Hartree–Fock equation for systems of up to 50 atoms. The complexity of the calculations scales like $n^4$. The Hartree-Fock approximation is equivalent to assuming that electrons move in an average field of all the other electrons in the system. The difference between the Hartree-Fock energy of a system and the exact energy is called the *correlation energy*. Improvements to the Hartree-Fock method, which are essential for most chemical applications, reintroduce in some approximate way the instantaneous repulsion between electrons and thus recover some of the correlation energy. The most common methods for improving Hartree-Fock include the configuration interaction method, many-body perturbation theory, and coupled cluster theory. Details on the Hartree-Fock method and its extensions are available in reference [4].

## 5.3 Introducing densities

Wavefunctions explicitly include the coordinates of all electrons (i.e., $\Psi = \Psi(\mathbf{r}_1, \ldots, \mathbf{r}_n)$). However, electrons are indistinguishable, and the position of a single electron in a molecule is not an observable. Thus we may ask: Can a theory be constructed that uses a simpler quantity, namely the charge density, as a fundamental quantity? The charge density is defined by

$$\rho(\mathbf{r}_1) = \int \cdots \int |\Psi(\mathbf{r}_1, \ldots, \mathbf{r}_n)|^2 d\mathbf{r}_2 \ldots d\mathbf{r}_n \ ,$$

and it integrates to $n$:

$$\int \rho(\mathbf{r}_1) d\mathbf{r}_1 = n \ .$$

The original idea of using charge density is embodied in the Thomas-Fermi-Dirac model of the atom, which is based upon the uniform electron gas. Suppose $n$ electrons are moving in a box of volume $V$ that contains a uniform distribution of positive charge that exactly neutralizes the electrons. Taking $n, V \to \infty$ as $\rho = n/V$ remains finite and constant, the allowable (infinite in number) energy states can be obtained. Assuming the state occupations obey a Fermi-Dirac distribution, one can show [1, pp. 47–49] that the kinetic and exchange interaction energies of the electrons in the uniform electron gas are

$$T(\rho) = C_F \int \rho^{5/3}(\mathbf{r}) d\mathbf{r} \;, \qquad C_F = \frac{3}{10}(3\pi^2)^{2/3} \;,$$

$$E_s(\rho) = -C_x \int \rho^{4/3}(\mathbf{r}) d\mathbf{r} \;, \quad C_x = \frac{3}{4}\left(\frac{3}{\pi}\right)^{1/3} \;.$$

Applying these expressions to an atom (a decidedly non-uniform system!) with nuclear charge $Z$ yields the Thomas-Fermi-Dirac model:

$$E_{TF}(\rho) = T(\rho) - Z \int \frac{\rho(\mathbf{r})}{r} d\mathbf{r} + \frac{1}{2} \int \int \frac{\rho(\mathbf{r}_1)\rho(\mathbf{r}_1)}{|\mathbf{r}_1 - r_2|} \, d\mathbf{r}_1 d\mathbf{r}_2 + E_x(\rho)$$

where the first and last terms are the uniform electron gas kinetic and exchange interaction energies, respectively, the second term is the nucleus-electron attraction, and the third term is the electron-electron repulsion term. Note that the exchange term also corrects for the spurious electron self-interaction energy in the electron-electron repulsion term.

In the ground state $\rho$ is the solution of the variational problem:

$$\begin{array}{ll} \text{minimize} & E_{TF}(\rho) \\ \text{subject to} & \rho \geq 0, \; \int \rho = n \;. \end{array} \tag{5.11}$$

While the Hartree-Fock method and its extensions provide reliable results for atomic and molecular systems, the Thomas-Fermi-Dirac model is much less satisfactory, primarily because of the highly approximate treatment of the electron kinetic energy. Nevertheless, the simplicity of the Thomas-Fermi-Dirac model encouraged other workers to incorporate some of the features of the electron density formulation into the Hartree-Fock method. Slater led the developments in this direction [5]. He rewrote the Hartree–Fock Coulomb operation as

$$\sum_j J_j(\mathbf{r}_1)\psi_i(\mathbf{r}_i) = \psi_i(\mathbf{r}_1)\left[\int \rho(\mathbf{r}_2)\frac{1}{r_{12}} \, d\mathbf{r}_2\right] = V_c[\rho(\mathbf{r}_1)]\psi_i(\mathbf{r}_1) \tag{5.12}$$

and the Hartree–Fock exchange operator as

$$\sum_j K_k(\mathbf{r}_1)\psi_i(\mathbf{r}_1) \int \frac{\rho_i^x(\mathbf{r}_2, \mathbf{r}_1)}{r_{12}} \, d\mathbf{r}_2 = V_x^i[\rho(r_1)]\psi_i(r_1)$$

where

$$\rho_i^x(\mathbf{r}_2, \mathbf{r}_1) = \sum \frac{\psi_i^*(\mathbf{r}_1)\psi_j^*(\mathbf{r}_2)\psi_j(\mathbf{r}_1)\psi_i(\mathbf{r}_2)}{\psi_i^*(r_1)\psi_i(\mathbf{r}_1)}$$

is the *exchange charge density.*

The exchange term may be viewed as producing an "exchange hole" around each electron. The exchange hole excludes electrons of the same spin from the neighborhood of an electron. The Hartree-Fock-Slater, or statistical exchange, treatment the potential due to this exchange hole is approximated by the corresponding expression for the uniform electron gas. The potential is thus obtained (up to a constant factor) as the Frechét derivative of $\int \rho^{5/3}(\mathbf{r})d\mathbf{r}$, i.e.,

$$V_x(\mathbf{r}_1) = -\frac{3}{2}\,\alpha\left(\frac{3\rho(\mathbf{r}_1)}{\pi}\right)^{1/3}$$

where $\alpha$ is a fudge factor ($\alpha \sim 0.7$). Then, instead of having the $n$ densities $\rho_i^x(\mathbf{r}_2, \mathbf{r}_1)$ in the Hartree-Fock equation, we have a simpler equation

$$f_x\Psi_i = \varepsilon_i\psi_i\ , \quad \text{where}$$
$$f_x = -\frac{1}{2}\,\nabla_1^2 - \frac{Z}{r_1} + V_c[\rho(\mathbf{r}_1)] + V_x(\mathbf{r}_1)\ . \tag{5.13}$$

This Hartree–Fock–Slater method is also called the $X\alpha$ method. It is a highly efficient model, and reasonably successful.

## 5.4 Density-functional theory

We seek a theoretical justification for the central roles of the charge density in the Thomas-Fermi-Dirac and Hartree-Fock-Slater models.

Consider a system of $n$ electrons moving in an external field $v(\mathbf{r})$, for instance the field generated by atomic nuclei:

$$v(\mathbf{r}) = \sum_\alpha \frac{Z_\alpha}{r_{1\alpha}}\ .$$

Since $v(\mathbf{r})$ and $n$ determine the Hamiltonian $\mathbf{H}$, they also determine the ground state $\Psi(\mathbf{r}_1, \ldots, \mathbf{r}_n)$ which is the minimizer of

$$\int \Psi^*\mathbf{H}\Psi\ .$$

The 1st Hohenberg–Kohn Theorem states that $v(\mathbf{r})$ is uniquely determined (up to an additive constant) by the density $\rho$ of the ground state, i.e., by

$$\rho(\mathbf{r}_1) = \int |\Psi|^2(\mathbf{r}_1, \mathbf{r}_2, \ldots, \mathbf{r}_n)d\mathbf{r}_2 \ldots d\mathbf{r}_n\ .$$

Consequently, the mapping

$$\Psi(\mathbf{r}_1, \mathbf{r}_2, \ldots, \mathbf{r}_n) \rightarrow \rho(\mathbf{r}_1) \tag{5.14}$$

is one-to-one. The question of characterizing the set of densities $\rho$ that correspond to potentials $v(\mathbf{r})$ is called the $v$-representability problem. This problem was studied in [6][7].

The 1st Hohenberg-Kohn theorem allows us to write the total energy of electrons in the field $v(\mathbf{r})$ *exactly* as:

$$E_v[\rho] = T[\rho] + V_{ee}[\rho] + \int \rho(\mathbf{r}_1)v(\mathbf{r}_1)d\mathbf{r}_1 \ ; \tag{5.15}$$

$T[\rho]$ is the kinetic energy, and $V_{ee}[\rho]$ is the electron interaction energy, i.e., the total energy is a unique functional of the charge density.

Because the proof of (5.14) (see [1, pp. 51–52]) is not constructive (the proof is by contradiction), the kinetic energy $T[\rho]$ and the electron interaction energy $V_{ee}[\rho]$, although they are universal functionals of $\rho$, have an unknown functional form!

The 2nd Hohenberg–Kohn Theorem asserts that $E_v[\rho]$ assumes its minimal value for the correct ground state $\rho(\mathbf{r})$ if the admissible functions $\rho$ are restricted by the condition

$$N[\rho] \equiv \int \rho(\mathbf{r}_1)d\mathbf{r}_1 = n \ . \tag{5.16}$$

Then, for the correct ground state $\rho$

$$E_0 \equiv E_v[\rho] \leq E_v[\tilde{\rho}] \quad \text{for any other acceptable} \quad \tilde{\rho} \ . \tag{5.17}$$

Equation (5.15) in the form presented is not amenable to solution by computation. If $V_{ee}[\rho]$ is entirely dropped, i.e. if we assume that the electrons do not interact, we can derive a workable expression for $E_v[\rho]$, namely,

$$E_v[\rho] = T_s[\rho] + \int \rho(\mathbf{r}_1)v(\mathbf{r}_1)d\mathbf{r}_1 \tag{5.18}$$

where $T_s[\rho]$ is the kinetic energy of the non-interacting system. Because the electrons do not interact, the wavefunction of the system can be expressed exactly as a sum over one-electron orbitals:

$$\rho(\mathbf{r}_1) = \sum_i |\psi_i(\mathbf{r}_1)|^2 \ . \tag{5.19}$$

With this substitution, $T_s[\rho]$ is given by

$$\begin{aligned} T_s[\rho] &= \int \Psi_s^* \left[ -\frac{1}{2} \sum_i \nabla_i^2 \right] \Psi_s d\mathbf{r}_1 \ldots d\mathbf{r}_n \\ &= -\int \frac{1}{2} \sum_i \psi_i^*(\mathbf{r}_1) \nabla_i^2 \psi_i(\mathbf{r}_1) d\mathbf{r}_1 \ . \end{aligned} \tag{5.20}$$

Substituting into (5.18) and applying the variational principle, we obtain

$$\left[-\frac{1}{2}\nabla_i^2 + v(\mathbf{r}_1)\right]\psi_i(\mathbf{r}_1) = \varepsilon_i\psi_i(\mathbf{r}_1) ,$$
$$E_0 = \Sigma\varepsilon_i .$$

Because the electron-electron interaction cannot be ignored, the above formulas (i.e., (5.18)–(5.20)) are incomplete. Kohn and Sham have subsequently constructed a correct form for $E_v[\rho]$ that resembles the special case of non-interacting electrons, and have thereby made the Hohenberg-Kohn approach more accessible to computations. They introduce

$$T_s[\rho] = \text{ non-interacting electron kinetic energy, as in (5.20),}$$

$$J[\rho] = \int \rho(\mathbf{r}_1)V_c(\mathbf{r}_1)d\mathbf{r}_1 = \text{ the Coulomb energy as in (5.12),}$$

and then *define* the exchange-correlation energy by

$$E_{xc}[\rho] = T[\rho] - T_s[\rho] + V_{ee}[\rho] - J[\rho]$$

where $T[\rho]$ is the kinetic energy of the true, interacting system.

Then (5.15) becomes

$$\begin{aligned} E_v[\rho] &= T_s[\rho] + J[\rho] + E_{xc}[\rho] + \textstyle\int \rho(\mathbf{r}_1)v(\mathbf{r}_1)d\mathbf{r}_1 \\ &\equiv T_s[\rho] + \textstyle\int \rho(\mathbf{r}_1)v_{\text{eff}}(\mathbf{r}_1)d\mathbf{r}_1 \end{aligned} \tag{5.21}$$

where the "effective potential" is given by

$$v_{\text{eff}} = V_c[\rho] + \frac{\delta E_{xc}}{\delta\rho} + v(\mathbf{r}_1) \qquad \left(V_c[\rho] = \frac{\delta J}{\delta\rho}\right) ; \tag{5.22}$$

here $\delta/\delta\rho$ denotes the Frechét derivative.

In calculating the ground state by the variational principle (5.17), the form (5.21) (with (5.22)) appears identical to that of the non-interacting electron system of the same density, but with a modified potential. Thus the variational principle yields

$$\begin{aligned} \left[-\frac{1}{2}\nabla^2 + v(\mathbf{r}_1) + V_c[\rho] + v_{xc}[\rho]\right]\psi_i(\mathbf{r}_1) &= \varepsilon_i\psi_i(\mathbf{r}_1) , \\ \text{where} \quad v_{xc}[\rho] &= \frac{\delta E_{xc}}{\delta\rho} ; \end{aligned} \tag{5.23}$$

$v_{xc}[\rho]$ is called the *exchange-correlation* potential. Equations (5.22), (5.23) together with (5.19) are called the Kohn–Sham (K–S) equations.

The exchange-correlation potential corrects for electron self-interaction, and produces an exchange-correlation hole about each electron. The form of the exchange-correlation potential is known only in special cases, but the

K–S equations can still be analyzed by scaling laws and in perturbation of special cases. The $\varepsilon_i$ in (5.23) are the smallest eigenvalues and the total energy of the system is

$$E = \sum_i \varepsilon_i - \frac{1}{2}\int \frac{\rho(\mathbf{r}_1)\rho(\mathbf{r}_2)}{|\mathbf{r}_1 - \mathbf{r}_2|}\, d\mathbf{r}_1 d\mathbf{r}_2 + E_{xc}[\rho] - \int v_{xc}(\mathbf{r}_1)\rho(\mathbf{r}_1)d\mathbf{r}_1;$$

we also have

$$\sum_i \varepsilon_i = T_s[\rho] + \int v_{\text{eff}}(\mathbf{r}_1)\rho(\mathbf{r}_1)d\mathbf{r}_1 \ .$$

Just as in the case of the Hartree–Fock treatment the total electronic energy is not the sum $\Sigma\varepsilon_i$ of the orbital energies.

It is important to note that the Hohenberg-Kohn theorems and the Kohn-Sham construction are only valid for the ground state of a system. However, the entire theory has an obvious connection with the methods that we have already discussed. Thus, the Hartree-Fock Slater method is obtained in the Kohn-Sham model by taking

$$v_{xc}[\rho] = -\frac{3}{2}\,\alpha\left(\frac{3}{\pi}\,\rho(\mathbf{r}_1)\right)^{1/3} \ .$$

i.e. the exchange potential of the uniform electron gas. The Hatree-Fock equation is similarly obtained by taking

$$v_{xc}[\rho] = -\sum_j K_j(\mathbf{r}_1) \ .$$

The so-called local density approximation is obtained by taking $v_{xc}$ as the exchange *and* electron correlation potentials of the uniform electron gas, the latter of which is available from accurate Monte Carlo simulations. Current developments are focused primarily on the "generalized gradient approximation," in which $v_{xc}$ is parameterized against $\rho$ and $\nabla\rho$ in such a way as to preserve correct scaling properties.

Current interest in the chemistry community in Density Functional Theory (DFT) arises from the development of accurate methods for numerically solving the Kohn-Sham equations, and the applicability to much larger systems than ab initio methods. The DFT suffers from several notable drawbacks which represent the cutting edge of modern research: the lack of systematic methods for improving $v_{xc}$, the restriction to electronic ground states, and the dependence on the Born-Oppenheimer approximation.

## 5.5 Applications

Bill Schneider presented three areas of his ongoing research in which DFT has or could find applications.

The first one has to do with Cu zeolites that are potential catalyst in catalytic converters. When hydrocarbon fuel burns up in the automobile engine, the following chemical reaction takes place:

$$\text{Fuel } (C_xH_y) + O_2 + N_2 \rightarrow CO_2 + H_2O + \text{ eneregy}$$
$$+ \text{ waste } HC, CO, NO.$$

The waste contains a collection of regulated pollutants. The catalytic converter converts the pollutants into relatively innocuous gases by the process:

$$HC + CO + NO + \text{ catalyst } \rightarrow CO_2 + H_2O + N_2 \ ;$$

for more details see [8, Chap. 7], [9, Chap. 21]. Standard catalysts are built to clean exhaust from a stoichiometric combustion mixture, which is the case in virtually all gasoline engines. However diesels and "lean-burn" engines which may be used in the future (because they potentially provide improved fuel economy) use a lean (excess $O_2$) mixture. Hence the need to build "lean-$NO_x$" catalysts to clean exhaust from lean mixtures. Cu zeolites, in particular Cu-ZSM-5, have the best known lean-$NO_x$ performance. Considerable research has been directed at understanding and improving these materials. Fundamental questions about the nature of the active sites in such catalysts and the precise catalytic mechanisms remain open. Recent computational work by W.F. Schneider, K.C. Hass, R. Ramprasad and J.B. Adams [10] uses DFT to examine models of the Cu-ZSM-5 catalyst.

The second application area deals with the atmospheric chemistry of HFCs, which are replacing CFCs in air conditioners and other applications. Figure 5.2 shows the fate of CFCs and HFCs in the atmosphere. CFCs are inert in the troposphere (0–5 miles altitude) but are degraded in the stratosphere (5–30 miles) to release chlorine atoms, which catalytically destroy ozone. HFCs, on the other hand, contain no chlorine and are thought to decompose primarily in the troposphere. Key questions that must be addressed include: how do HFCs decompose?; what trace species do HFCs release in the stratosphere?; do the trace species impact stratospheric ozone? A summary of these issues is provided in reference [11]. Ab initio calculations have been valuable in addressing these questions. Unfortunately, because of the limitations mentioned above, DFT is not well-suited to these problems. An example of these limitations, dealing with the breakdown of the Born-Oppenheimer approximation, is given in reference [11].

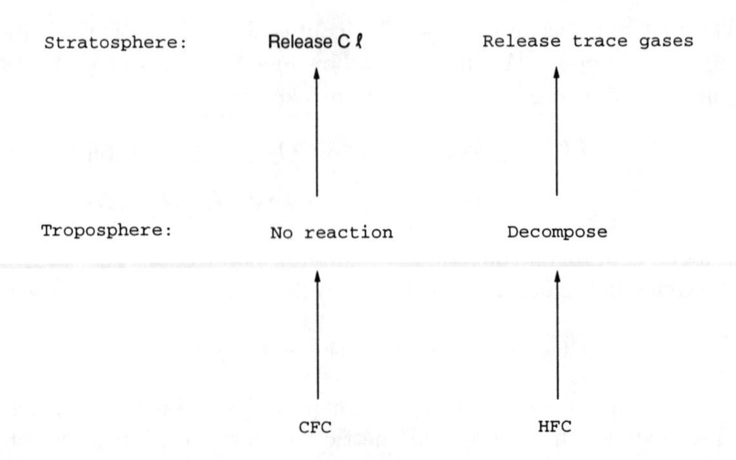


FIGURE 5.2.

The third area of application deals with paint chemistry. The automobile outer panels are painted with basecoats and finished with clearcoat that contains ultraviolet absorbers (UVAs) that protect the basecoat from sunlight. When the automobile is in the sun, photons from the sunlight collide with the UVAs as illustrated in Figure 5.3

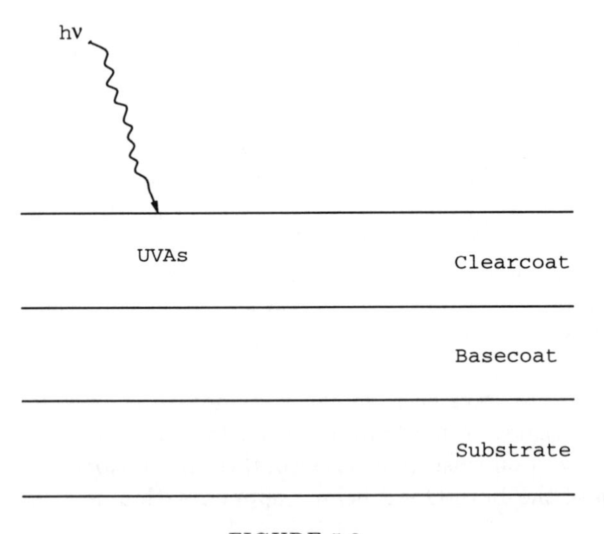


FIGURE 5.3.

The photon $h\nu$ causes the UVA molecule to go into an excited state, as shown in Figure 5.4 where [ ]* indicates the molecule in excited state. The energy dissipates as thermal energy and the excited molecule returns to its ground state.

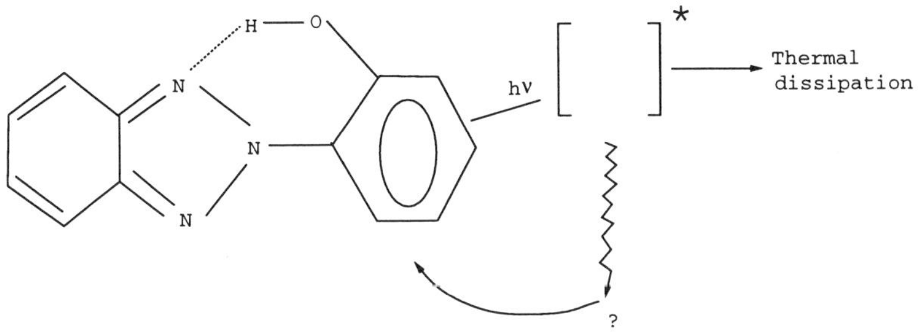

FIGURE 5.4.

The Hohenberg–Kohn theorems are valid only for non-degenerate ground states. Thus the density functional theory (DFT) (with its Kohn–Sham development) does not apply to excited states. The example with the UVAs described above cannot therefore be treated by the DFT. UVA photochemistry is again modeled primarily with ab initio methods.

There is a need to extend the DFT method beyond the ground state. Several ad hoc methods have been proposed, such as time-dependent DFT and Slater half-electron method. Developing a rigorous theory of excited states within DFT is a particularly challenging problem.

## 5.6 References

[1] R.G. Parr and W. Yang, *Density-Functional Theory of Atoms and Molecules*, Oxford University Press, New York (1989).

[2] N.H. March, *Electron Density Theory of Atoms and Molecules*, Academic Press, New York (1989).

[3] L.J. Sham and M. Schlinter, *Principles and Applications of Density Functional Theory*, Word Scientific Teaneck, N.J. (1989).

[4] A. Szabo and N.S. Ostlund, *Modern Quantum Chemistry*, McGraw-Hill, New York (1989).

[5] J.C. Slater, *Statistical Exchange-Correlation in the Self-Consistent Field*, Adv. Quantum Chem. 6 (1972), 1.

[6] J.T. Chayes, L. Chayes, and E.H. Lieb, *The inverse problem in classical statistical mechanics*, Commun. Math. Physics, 93 (1984), 57–121.

[7] J.T. Chayes and L. Chayes, *On the validity of the inverse conjecture in classical density functional theory*, J. Stat. Physics, 36 (1984), 471–488.

[8] A. Friedman, *Mathematics in Industrial Problems, Part 4*, IMA Volume 38, Springer-Verlag, New York (1991).

[9] A. Friedman, *Mathematics in Industrial Problems, Part 6*, IMA volume 57, Springer–Verlag, New York (1994).

[10] W.F. Schneider, K.C. Hass, R. Ramprasad and J.B. Adams, *Cluster models of Cu binding and CO and NO adsorption in Cu-Exchanged zeolites.* Preprint (1995).

[11] T.J. Wallington, W.F. Schneider, D.R. Worsnop, O.J. Nielsen, J. Sehested, W.J. DeBruyn, and J.A. Shorter, *The environmental Impact of CFC Replacements—HFCs and HCFCs*, Environ. Sci. Technol., 28 (1994), 320A–326A.

[12] W.F. Schneider, M.M. Maricq, and J.S. Francisco, *The vibrational spectrum of FC(O)O radical: a challenging case for single-reference electron correlation methods*, J.Chem. Phys., 103 (1995), 6601–6607.

# 6

# Nonlinear diffusion in coating flows

Liquid curtains issuing from a slot and falling under the influence of gravity are found in a wide variety of industrial processes. In particular this technique is employed to deposit a thin uniform film on a moving solid surface during coating process [1]. The fluid sheet tends to break because small holes and streaks in the curtain spread and grow due to surface tension [2]. One way to stabilize the coating flow is to use surfactants, surface-active chemicals (such as detergents), that reduce surface tension. Their effectiveness depends on their chemical properties and on the interaction with the flow of the fluid. On December 1, 1995 David Ross from Eastman Kodak described the basic fluid dynamics of curtain coating flows and of the basic chemistry of surfactants. He introduced a simple nonlinear diffusion-convection model of the flow of surfactants in a curtain, and discussed the characteristics of a good surfactant; his model is aimed at helping identify optimal surfactants. Finally he presented open problems which arise in measuring surface tension in curtain coating.

## 6.1 Curtain coating

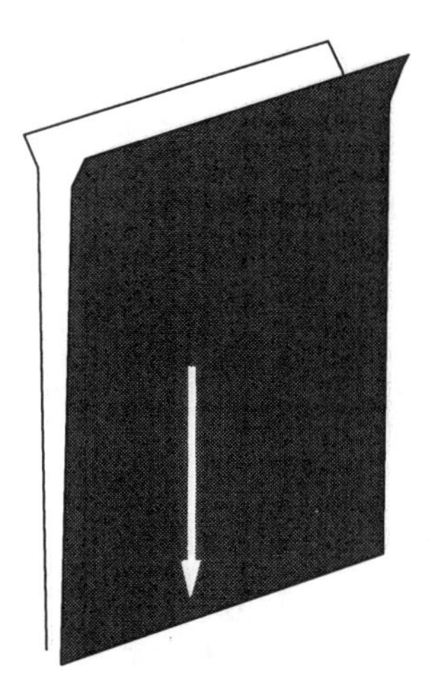

FIGURE 6.1.

In curtain coating, fluid is extruded from a slot in a hopper. Figure 6.1 shows a nearly front view of the curtain flow. In order to keep the edges of the curtain from narrowing down as the flow keeps falling, a problem known as edge withdrawl, vertical wires (edge guides) are placed at the edges of the curtain; surface tension then keeps the curtain in contact with the wire, and so the curtain flow maintains its rectangular front view profile between the bounding wires.

Figure 6.2 shows side view of curtain coating. The thickness of the fluid layer decreases downward since its velocity increases; it is essentially in gravitational free fall.

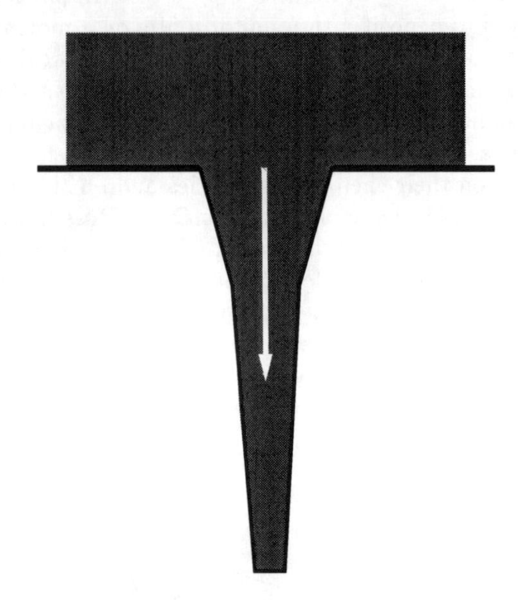

FIGURE 6.2.

If water is thinly spread on a flat surface, small perturbations will result in the formation of isolated drops; this is due to surface tension. In the same way surface tension acts to increase a small dent in the curtain, to spread and magnify streaks, and to split the curtain flow; this will result in non-uniform coating, which is undesirable. To counter this we add surfactants to the fluid. Surfactants are hydrophobic species; they like to migrate from the bulk of the fluid to the surface. When they adsorb at the surface they cause reduction in the surface tension. As the flow keeps falling down, more surfactant reaches the fluid boundary, and the surface tension decreases; we have a situation of *dynamic* surface tension (DST).

The general question is: How best to use surfactants to achieve stability of curtain coating?

The surfactants we consider here are negatively charged ions (i.e., they are anionic surfactants). It is convenient to visualize them as having head and tail:

The tail is hydrophobic: it does not like water. The longer the tail, the more hydrophobic the species is. The species is characterized by the length of the tail; a homologous family is a class of surfactants, all of which have the same head group and which differ only in tail length. It might appear that the DST reduction should increase monotonically with the length tail $\tau$. Yet a surprising experiment by Alan Pitt (see [3]) shows that in a homologous family the optimal DST reduction occurs at some length $\tau_0$, and for $\tau > \tau_0$ it goes up again; see Figure 6.3.

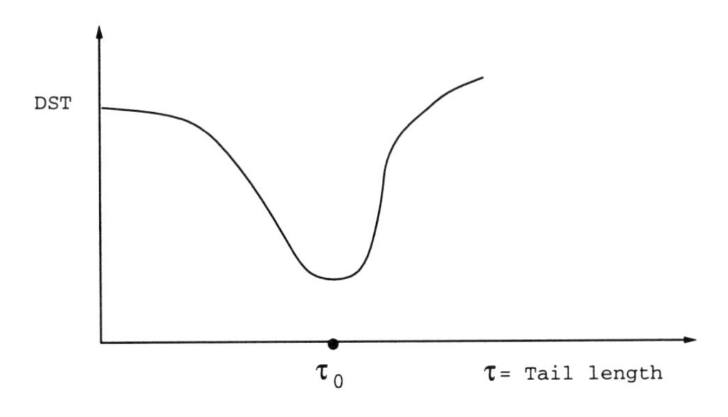


FIGURE 6.3.

Another feature of surfactants was invoked to explain this phenomenon [3]. At sufficiently high concentrations, surfactants will form micelles, small clusters of surfactants molecules; see Figure 6.4. In a micelle, the tails of the molecules point toward the interior of the cluster, and the heads face the aqueous solution. Surfactants, because of their hydrophobicity, 'prefer' such an arrangement; in a micelle, the surfactants' hydrophobic tails are packed together, shielded from the water. Micelles are typically composed of 60 or more monomers. For any surfactant there is a well-defined concentration, the critical micelle concentration, or CMC, above which micelles form. Any concentration above the CMC goes into solution in the form of micelles.

We take the CMC as a measure of the hydrophobicity of a surfactant, within a homologous family. As the tail length increases—the hydrophobicity increases—the CMC decreases; the more hydrophobic the species, the more 'eager' it is to form micelles.

FIGURE 6.4.

Micelles are much larger than monomer, so they diffuse slowly. As CMC decreases—the surfactants get more hydrophobic—the monomer is more surface active, it causes greater surface tension reduction. However, as CMC decreases, more surfactant gets tied up in slowly-diffusing micelles, so less of it gets to the surface. The model expresses this trade-off mathematically.

We introduce the quantities:

$y$ = distance from the interface (see Figure 6.5),

$u(y,t)$ = unimer concentration,

$m(y,t)$ = micelle concentration,

$z(y,t) = u(y,t) + Nm(y,t)$ = total concentration,

where $N$ = number of unimers per micelle,

$\Gamma(t)$ = surface concentration of unimers (at the liquid-air interface).

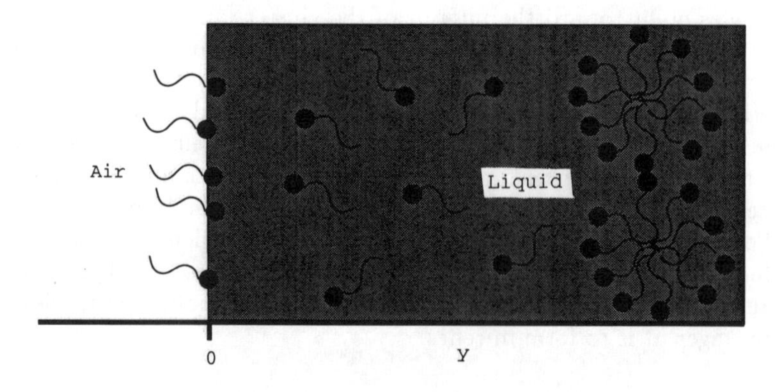


FIGURE 6.5.

Then

$$\frac{\partial u}{\partial t} = D_u \frac{\partial^2 u}{\partial y^2} - f(u)N \ ,$$

$$\frac{\partial m}{\partial t} = D_m \frac{\partial^2 m}{\partial y^2} + f(u)N$$

where $D_u$ and $D_m$ are the diffusion coefficients of unimers and micelles ($D_u > D_m$). The function $f(u)$ has the form shown in Figure 6.6.

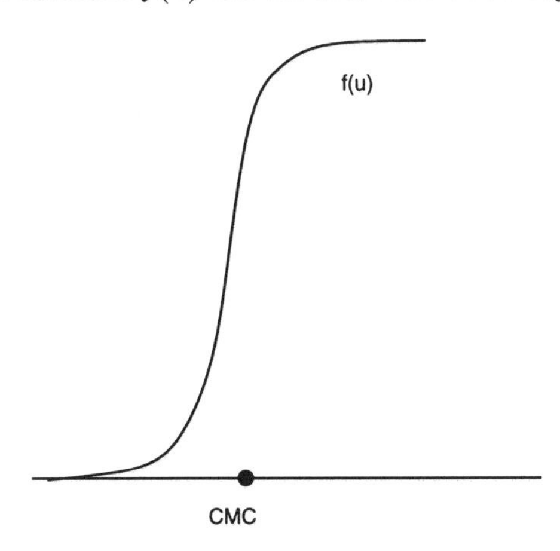


FIGURE 6.6.

We shall assume for simplicity that $f(u) = 0$ if $u < CMC$ and that the concentration of $u$ never exceeds CMC; all surfactant concentration above the CMC goes into micelles. Then, at any point in space, either the micelle diffusion term is 0—because the total concentration is below CMC, so there are no micelles—or the unimer diffusion term is 0—because the total concentration is above CMC, so the unimer concentration is identically equal to the CMC. Thus, we can combine the last two equations into a single equation by adding, to get

$$\frac{\partial z}{\partial t} = \frac{\partial}{\partial y}\left(D(z)\frac{\partial z}{\partial y}\right) \quad \text{if} \quad 0 < y < \infty, \ t > 0 \tag{6.1}$$

where

$$D(z) = \begin{cases} D_u & \text{if} \quad z \leq CMC \\ D_m & \text{if} \quad z > CMC \ . \end{cases} \tag{6.2}$$

To formulate the boundary condition at $y = 0$, set

$$\Gamma_{\max} = \text{maximum surface concentration of unimers.}$$

We make two assumptions:

(i) The flux of $z$ from the bulk into the air-liquid interface results in unimers only, so that

$$\frac{d\Gamma}{dt} = D(z)\frac{\partial z}{\partial y} \quad \text{at} \quad y = 0 \quad \text{provided } z < CMC\ ;$$

$$\text{if however} \quad z \geq CMC \quad \text{then} \quad \frac{\partial z}{\partial y} = 0\ .$$

(ii) At the interface the unimer in solution is in equilibrium with the unimer on the surface. This equilibrium is described by a Langmuir isotherm

$$z(0,t) = \alpha \cdot CMC \frac{\Gamma(t)}{\Gamma_{\max} - \Gamma}$$

where $\alpha$ is a positive constant.

If we differentiate the last equation in $t$ and use the equation from (i) (for $z < CMC$), we get

$$\frac{\partial z(0,t)}{\partial t} = \frac{\alpha \cdot CMC}{\Gamma_{\max}} \left(1 + \frac{z(0,t)}{\alpha \cdot CMC}\right)^2 D(z(0,t))\frac{\partial z(0,t)}{\partial y} \quad \text{if } z < CMC\ . \tag{6.3}$$

Recall also that

$$\frac{\partial z(0,t)}{\partial y} = 0 \quad \text{if} \quad z \geq CMC\ . \tag{6.4}$$

Finally we prescribe an initial condition

$$z(y,0) = b\ . \tag{6.5}$$

One can solve the system (6.1)–(6.5) and then use the solution to compute the surface tension $\gamma$ by Frumkin's equation:

$$\gamma = \gamma_0 - RT\Gamma_{\max} \log \frac{1}{1 - \dfrac{\Gamma(t)}{\Gamma_{\max}}} = \gamma_0 - RT\Gamma_{\max} \log\left(1 + \frac{1}{\alpha}\frac{z(0,t)}{CMC}\right)\ .$$

The above mathematical model was set up by Pitt, Ross and Whitesides [3]. Fixing time $t$, they computed the surface tension as a function of CMC within a homologous family; the CMC varies from one homologous family of surfactants to another. Figure 6.7 shows a very good fit to experiment; note that the model contains no free parameters. Excellent fit to experiment was also established (in [3]) for $\gamma$ as a function of the concentration.

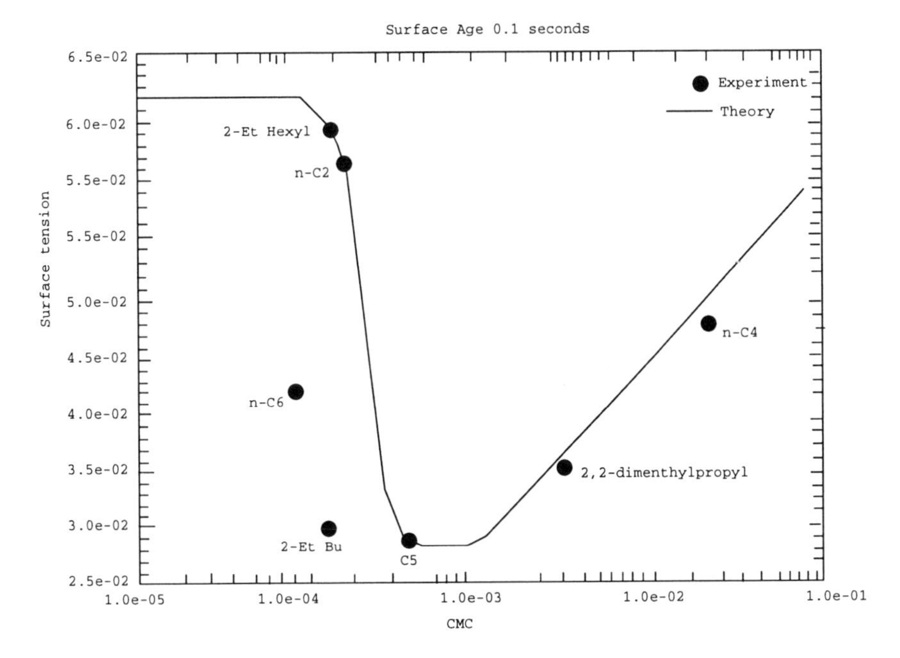


FIGURE 6.7.

The above model did not take into account two velocities:

$u(t)$ = vertical velocity,

$c(y,t)$ = net velocity toward the interface (this velocity is in the horizontal direction).

Taking these velocities into account, (6.1) is replaced by

$$\frac{\partial z}{\partial t} = \frac{\partial}{\partial y}\left(D(z)\frac{\partial z}{\partial y}\right) - c(y,t)\frac{\partial z}{\partial y} ,$$

and (6.3) is replaced by

$$\frac{\partial z}{\partial t} = \frac{\alpha \cdot CMC}{\Gamma_{\max}}\left(1+\frac{1}{\alpha \cdot CMC}\right)^2 D(z(0,t))\frac{\partial z}{\partial y} - z\left(1+\frac{z}{\alpha \cdot CMC}\right)\frac{(du/dt)}{u} \quad \text{at } y=0 .$$

## 6.2 Measuring DST

The measurement of the surface tension $\gamma$ as reported in [3] was done in an experiment where the fluid is not accelerated. In the curtain coating, however, the fluid is accelerated and it is therefore desirable to develop techniques for measuring dynamic surface tension (DST). If we make a shallow dent on the surface of a curtain, this develops into a streak or a ripple on the coating surface, as illustrated in Figure 6.8.

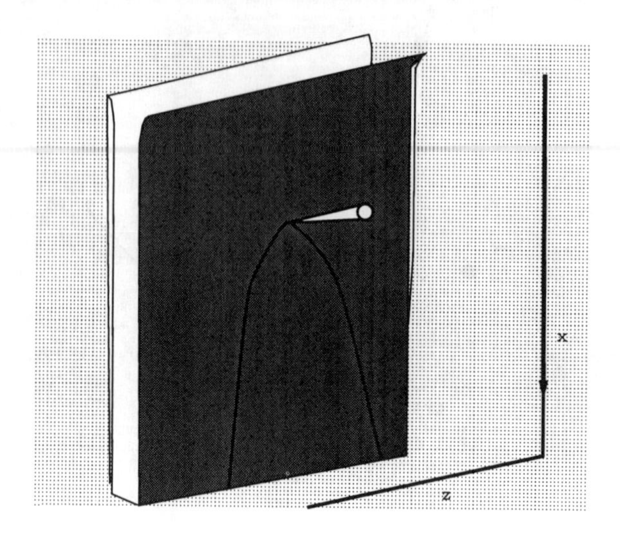


FIGURE 6.8.

According to Lin and Roberts [4] the ripple is given by a curve $x = x(z)$ such that

$$\frac{dz}{dx} = \pm\sqrt{\frac{1}{\frac{\rho q u}{2\gamma} - 1}} \tag{6.6}$$

where

$q =$ mass flux in the curtain,

$u =$ speed of fluid in the ripple,

$\rho =$ density of fluid, and

$\gamma =$ surface tension.

To carry out the experiment shown in Figure 6.8 requires some care; if one is not too careful, the curtain will break with edge bead, as illustrated in Figure 6.9. Since it is very easy to break curtains with edge bead and then to measure the resulting bead shape, the question arises: How does the bead shape depend on surface tension?

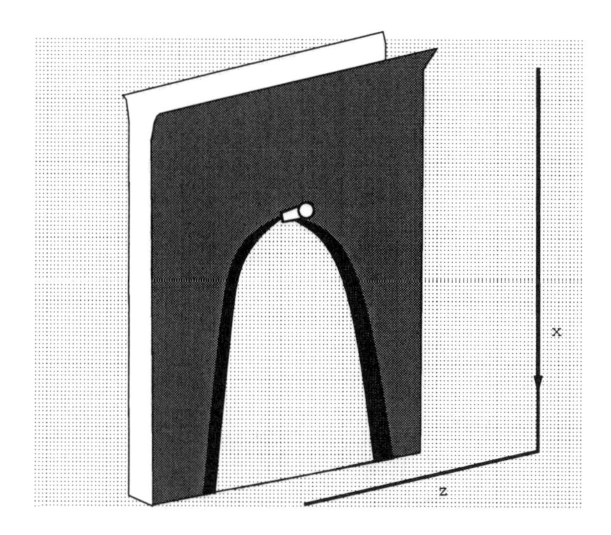


FIGURE 6.9.

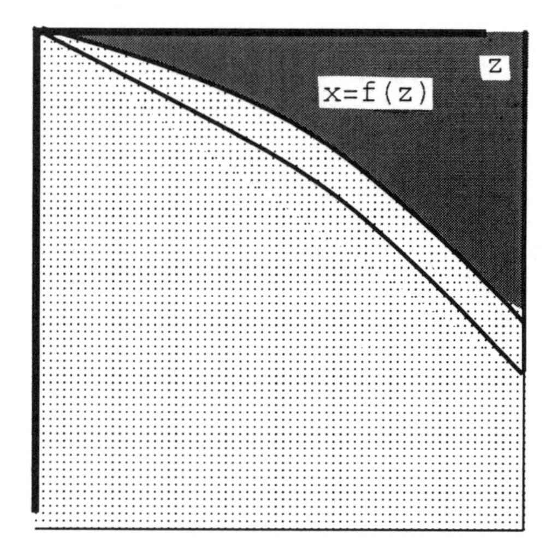


FIGURE 6.10.

The answer was given by Baumlin, Ross, Weinstein and Whitesides [5]. Denote by $x = f(z)$ the edge bead (which is actually a tube of fluid; see Figure 6.10). Set

$M(z)$ = mass per unit length of the bead,

$\nu(z)$ = fluid speed in the bead,

$q$ = mass flux in the curtain,

$g$ = gravitational constant,

$u_0$ = fluid velocity at the top of the curtain,

$u(x) = \sqrt{u_0^2 + 2gx}$ = fluid speed in the curtain (free fall).

Figure 6.11 describes the fluid flow at the boundaries of a short element of the bead, and the forces acting on this element (i.e., surface tension and gravity). The radius of tube is several times larger than the thickness of the curtain.

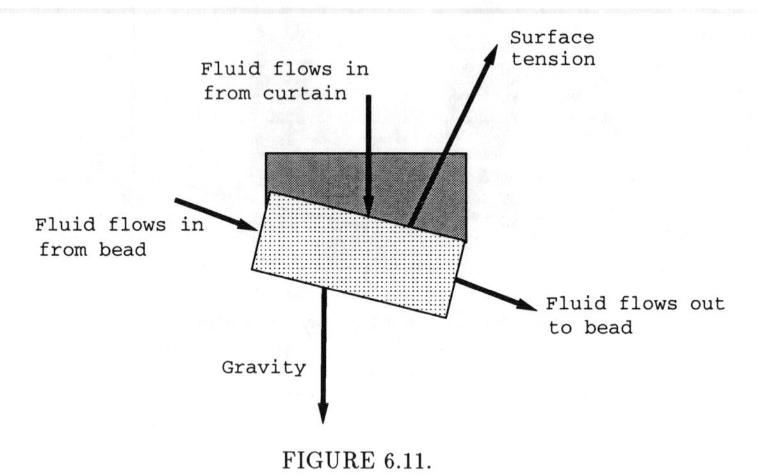


FIGURE 6.11.

By conservation of mass (the mass of fluid entering the element from above and from the left cross-section is equal to the amount of fluid which exits from the right cross-section):

$$\frac{d}{dz}(M(z)\nu(z)) = \rho u(x) = q$$

where $\rho$ is the density of the fluid in the curtain and $q$ is assumed to be constant. Hence, if we also assume that

$$M(0) = 0 , \tag{6.7}$$

and choose the origin of the $(x, z)$-coordinate system so that $f(0) = 0$, then

$$M\nu = qz . \tag{6.8}$$

The conservation of momentum gives:

$$-\int \rho v_i v_j n_j dA + \int F_i \rho dV + \int \sigma_{ij} n_j dA = 0 \tag{6.9}$$

where $(F_i)$ is the gravitational force, $(v_i)$ the fluid velocity, $\sigma_{ij}$ the stress tensor, $(n_i)$ the outward normal, the volume integration is over the volume $V$ of the tubular element, and the surface integration is over its boundary.

We note that the balance of forces on the fluid/air interface gives $\sigma_{ij}n_j = \gamma n_i$; we shall henceforth assume that $\gamma$ is constant.

By (6.9) in the vertical $x$-direction we get

$$-\rho u^2 + \frac{d}{dz}\left(M(z)(\nu(z))^2 \frac{f_z}{\sqrt{1+f_z^2}}\right) = gM(z)\sqrt{1+f_z^2} - 2\gamma \qquad (6.10)$$

where $f_z$ means the derivative $df/dz$. Similarly, by (6.9) in the horizontal $z$-direction we get

$$\frac{d}{dz}\left(\frac{M(z)(\nu(z))^2}{\sqrt{1+f_z^2}}\right) = 2\gamma f_z$$

and, by integrating and using (6.7),

$$\frac{M\nu^2}{\sqrt{1+f_z^2}} = 2\gamma f \ . \qquad (6.11)$$

Note that the reason we have $2\gamma$ (instead of $\gamma$) is that both sides of sheet are pulling the element of the bid by surface tension. From (6.8), (6.11) we obtain the relations

$$\frac{qz\nu}{\sqrt{1+f_z^2}} = 2\gamma f \ ,$$

and

$$gM\sqrt{1+f_z^2} = g\frac{qz}{\nu}\ \sqrt{1+f_z^2} = g\frac{q^2}{2\gamma f}\ z^2 \ .$$

Using these relation in (6.10) yields an ODE for $f$:

$$f^2\frac{d^2 f}{dz^2} + f\left[\left(\frac{df}{dz}\right)^2 + 1 - \frac{qu_0}{2\gamma}\left(1 + \frac{2gf}{u_0^2}\right)^{1/2}\right] = g\left(\frac{q}{2\gamma}\right)^2 z^2 \ . \qquad (6.12)$$

We have assumed that $f(0) = 0$. To solve (6.12) we need another condition at $z = 0$. We shall require that

$$f(z) \text{ is analytic at } z = 0, \text{ so that} \qquad (6.13)$$
$$f(z) = a_1 z + \frac{a_2}{2!}\ z^2 + \cdots \text{ near } z = 0 \ .$$

Then we get

$$a_1 = \left(\frac{qu_0}{2} - 1\right)^{1/2}$$

which corresponds to the formula (6.6).

Figure 6.12 shows the computational and experimental results for both small disturbance (Figure 6.8) and for the bead formation (Figure 6.9)).

Although the assumptions (6.7), (6.3) for the edge computation have not been justified on physical grounds, there is good fit with the actual data.

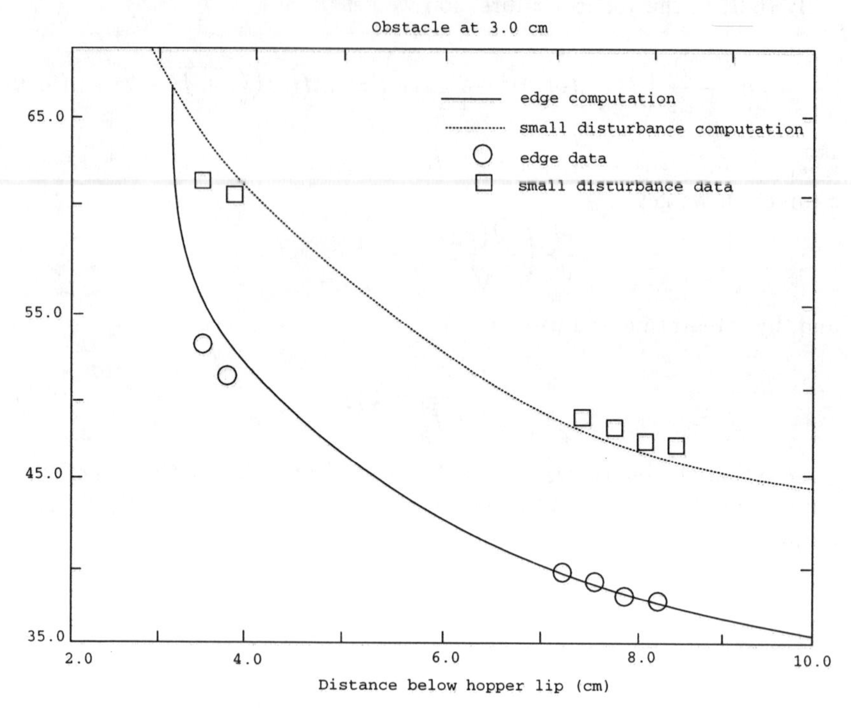


FIGURE 6.12.

# 6.3 Open Problems

*Problem (1).* What does the flow look like near the obstacle that breaks the curtain (in Figure 6.9)?

One has to consider here the Navier–Stokes equations, with the free boundary conditions: (i) surface tension balances the viscous stress at the free boundary; (ii) on the boundary of the solid obstacle the no-slip condition holds, and (iii) contact angle condition should be imposed where the free boundary meets the surface of the obstacle.

*Problem (2).* Derive asymptotic formulas for the previous problem in case the width $\varepsilon$ of the curtain coating goes to zero.

An informal argument in the absence of surface tension was given by G. Taylor in the Appendix to [6]. G. Taylor [2] has also considered the case of surface tension provided the Weber number $W = \dfrac{2\gamma}{\rho q u}$ is $< 1$.

The condition $W < 1$ ensures stability of the curtain. (In the experiments reported in [3], $\gamma \sim 70$ dynes/cm, the curtain thickness is $\frac{1}{2}$ mm and the free fall velocity of the fluid is $u \sim m$/sec; in this case $W \sim 0.07$.) We refer to [1] for more recent results on the stability of liquid curtains.

## 6.4 References

[1] D.S. Finnicum, S.J. Weinstein and K.J. Ruschak, *The effect of applied pressure on the shape of a two-dimensional liquid curtain falling under the influence of gravity*, J. Fluid Mech., 255 (1993), 647–665.

[2] G. Taylor, *The dynamics of thin sheet of fluid III. Disintegration of fluid sheets*, Proc. Royal Soc., A, 253 (1959), 313–321.

[3] A. Pitt, D. Ross and T. Whitesides, *A mathematical model of dynamic surface tension*, submitted to Langmuir.

[4] S.R. Lin and G. Roberts, *Waves in a viscous liquid curtain*, J. Fluid Mech., 112 (1981), 443–458.

[5] Baumlin, D. Ross, S. Weinstein and T. Whitesides, in preparation.

[6] D.R. Brown, *A study of the behavior of thin sheet of moving liquid*, J. Fluid Mech., 10 (1961), 297–305.

# 7

# A mini-max pursuit evasion algorithm

The aerospace industry is interested in 3D mini-max pursuit-evasion algorithms that can run in real-time on currently available missile-control computers. In today's technology, fighter aircraft can evade anti-aircraft missiles. This is due either to sub-optimal guidance algorithms or to algorithms with slow execution time. On January 5, 1996 Michael Elgersma from Honeywell Technology Center described a real time 3D mini-max pursuit-evasion algorithm that he and Blaise Morton (also from Honeywell) have recently developed [1]. This algorithm allows unknown intercept time, assumes that the aircraft and missile can each accelerate perpendicular to their velocity, and that the aircraft can turn in a way to best avoid the missile and do it under realistic physical constraints.

## 7.1 The problem

A highly maneuverable fighter aircraft can evade anti-aircraft missiles. Part of the problem is due to sub-optimal guidance algorithms. An optimal algorithm would be game theoretic: the missile maneuvers to intercept the aircraft while the aircraft maneuvers to avoid the missile. If each is basing its moves in accordance with the information available on the past behavior and present position, state and velocity of the other (but with no information about future intentions), then we are within the framework of differential games [2][3].

An optimal algorithm would account for nonlinearities in the dynamics and the nonlinearities in the acceleration constraints.

Off-line algorithms can be used to find optimal strategies, but they essentially do grid search on high (4 or 7) dimensional spaces so they cannot be used in real-time applications. The 4-dimensional spaces are based on constant accelerations perpendicular to the motion of the moving body: 2 for the missile and 2 for the aircraft; the 7-dimensional spaces are based on variable 3D accelerations (3 missile and 3 aircraft accelerations, plus time).

Current real-time guidance algorithms make some of the following assumptions:

(i) intercept time is known;

(ii) aircraft is not maneuvering at all;

(iii) aircraft maneuvering in random way modelled by Brownian motion;

(iv) linear dynamics;

(v) sensor noise is zero (fuel optimal control, bang bang).

The objective here is to develop a real-time algorithm that will allow the following:

(i) unknown intercept time;

(ii) aircraft and missile can each accelerate perpendicular to their velocity;

(iii) aircraft can turn in a way that best avoids the missile;

(iv) linear point mass dynamics is assumed, but acceleration constraints turn with vehicle, and

(v) acceleration is a smooth function of sensor input.

The standard approach is to grid 2 components of missile acceleration perpendicular to its velocity, and 2 components of the aircraft acceleration perpendicular to its velocity. Then compute the cost functional for each resulting set of trajectories, and mini-max it.

If we denote by $T_a(\mathbf{r})$ the aircraft arrival time at a point $\mathbf{r}$, and by $T_m(\mathbf{r})$ the missile arrival time of the point $\mathbf{r}$, then a point $\mathbf{r}_0$ is on the *intercept set* if $T_a(\mathbf{r}_0) = T_m(\mathbf{r}_0)$. If both aircraft and missile choose constant accelerations perpendicular to the velocity, then the *intercept set* is a 2-dimensional surface in the 3D space. The new approach developed in [1] grids the 2-dimensional intercept surface and for each $\mathbf{r}$ on this surface computes the constant accelerations that make both the aircraft and missile hit the point $\mathbf{r}$ at the same arrival time. It then computes the cost functional for the resulting set of trajectories and proceeds to execute the mini-max on this functional, as $\mathbf{r}$ varies on the intercept set. This approach will be explained in more details in the next two sections.

## 7.2 The intercept surface

Let vehicle 1 be the evader (aircraft) and vehicle 2 be the pursuer (missile). Denote by $\mathbf{r}_i(t)$ the position of vehicle $i$ at time $t$, and by $\mathbf{V}_i(t)$ its velocity at time $t$. Let $m_i$ denote the mass of vehicle $i$, and let $\mathbf{u}_i$ denote a perpendicular force on vehicle $i$; we can parametrize $\mathbf{u}_i$ so that it belongs to a set $U_i$ in $\mathbb{R}^2$.

We recall that inertial coordinates $(x, y, z)$ are coordinates for which Newton's law

$$\text{Force} = (\text{mass}) \cdot (\text{acceleration})$$

is valid. For rigid body, such as in Figure 7.1, it is also convenient to introduce body axes coordinates $(\tilde{x}, \tilde{y}, \tilde{z})$; this is an orthonormal system with center fixed in the body, and with the directions of the coordinate axes marked in the body. We can take for example the origin to be at the position of the pilot, the $\tilde{x}$-axis in the direction of the long axis of the aircraft, and the $\tilde{y}$-axis in a direction parallel to the wings; see Figure 7.1.

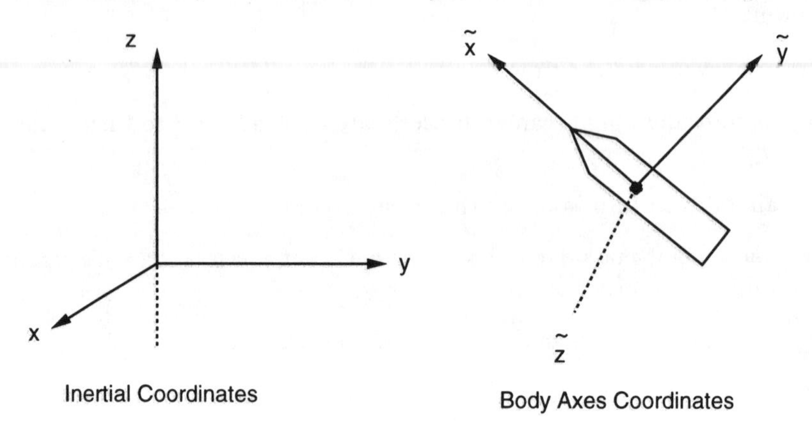


FIGURE 7.1.

Let $\mathbf{V}_i^{\perp}(t)$ be an orthonormal complement of $\mathbf{V}_i(t)$; it can be realized as a $3 \times 2$ matrix that varies with $t$. Then, the constant-speed, point-mass equations of motion (i.e., the equations for the center of mass) in inertial coordinate are given by

$$\begin{aligned} m_i \dot{\mathbf{V}}_i &= \mathbf{V}_i^{\perp}(t) \mathbf{u}_i \ , \\ \dot{\mathbf{r}}_i &= \mathbf{V}_i(t) \end{aligned} \tag{7.1}$$

with initial data

$$\mathbf{V}_i(0) \ , \ \mathbf{r}_i(0) \ . \tag{7.2}$$

Note that $\mathbf{u}_i$ is fixed with respect to the body axes coordinates, and $\mathbf{V}_i^{\perp}(t)$ rotates the vehicle (in the inertial coordinates).

The following well known kinematic equation holds:

$$\frac{d}{dt} \begin{pmatrix} \dot{x} \\ \dot{y} \\ \dot{z} \end{pmatrix} = \boldsymbol{\omega} \times \begin{pmatrix} \dot{x} \\ \dot{y} \\ \dot{z} \end{pmatrix} + \text{acceleration} \tag{7.3}$$

where $\boldsymbol{\omega}$ is a vector accounting for the rotation of the body. If the acceleration is perpendicular to the velocity vector $\mathbf{V}$, as in (7.1), then taking the scalar product of $\mathbf{V} = (\dot{x}, \dot{y}, \dot{z})^T$ with (7.3) we find that

$$\frac{d}{dt}(\dot{x}^2 + \dot{y}^2 + \dot{z}^2) = 0 \ ,$$

that is, the body maintains constant speed.

We assume that the constant normal acceleration motion, as defined in (7.1), is chosen so that the point mass moves along a circular path in the plane spanned by $\mathbf{V}_i(0)$ and $\dot{\mathbf{V}}_i(0)$.

The set of constant-speed circular trajectories through the point $\mathbf{r}_i(0)$ with initial velocity $\mathbf{V}_i(0)$ is 2-dimensional and can be parametrized by the two components of the normal force $\mathbf{u}_i$. The constraint that the circles for which vehicle 1 and vehicle 2 intersect reduces the 4 dimensional set to a 3-dimensional set. The constraint that both vehicles arrive at the intercept point at the same time, further reduces the intercept surface to a 2-dimensional set. Figure 7.2 shows typical intercepts.

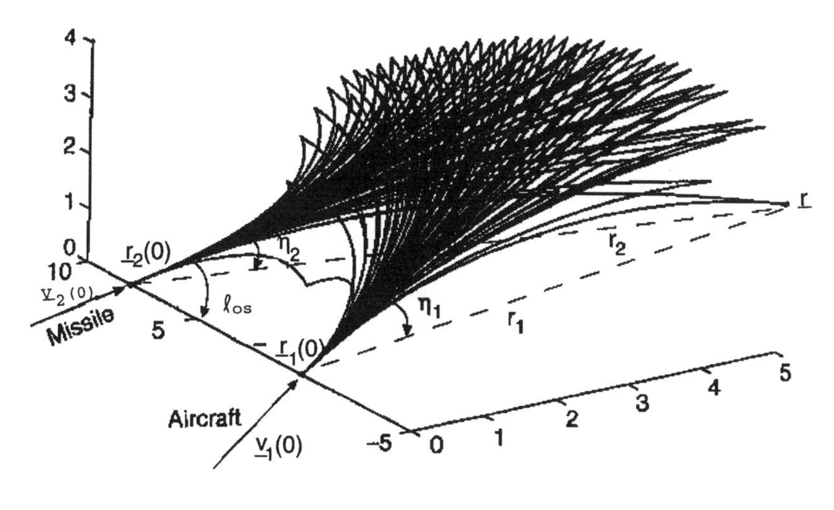


FIGURE 7.2.

The intercept surface separates the 3D space into regions where the aircraft can get to first and regions where the missile can get to first.

We proceed to compute the intercept surface.

We define two sets of polar coordinates $(r_i, \eta_i, \xi_i)$ based at $\mathbf{r}_i(0)$ by

$$\mathbf{r} = \mathbf{r}_i(0) + r_1 \left[ \frac{\mathbf{V}_i(0)}{||\mathbf{V}_i(0)||} , \ \mathbf{V}_i(0)^{\perp} \right] \begin{bmatrix} \cos \eta_i \\ \sin \eta_i \cos \xi_i \\ \sin \eta_i \sin \xi_i \end{bmatrix} . \tag{7.4}$$

Thus

$$r_i = ||\mathbf{r} - \mathbf{r}_i(0)|| \ , \ \eta_i = \cos^{-1} \left[ \frac{\mathbf{V}_i(0) \cdot (\mathbf{r} - \mathbf{r}_i(0))}{||\mathbf{V}_i(0)|| r_i} \right] .$$

Figure 7.3 shows the arc travelled by vehicle $i$ from time $t = 0$ to time $T_i$, when the vehicle has reached point $\mathbf{r}(T_i)$. Since vehicle $i$ travels with constant speed $||\mathbf{V}_i(0)||$, the arc length is $||\mathbf{V}_i(0)||T_i$. On the other hand, observe that the vehicle has rotated by angle $2\eta_i$ and the bisector $AB$ goes through the center of the circle traveled by vehicle $i$. Hence the ratio of the circular distance to the chord distance from $\mathbf{r}_i(0)$ to $\mathbf{r}(T_i)$ is $\eta_i/(\sin\eta_i)$, and thus

$$T_i = \frac{r_i}{||\mathbf{V}_i(0)||}\frac{\eta_i}{\sin\eta_i}$$

The intercept surface is obtained by solving for all $\mathbf{r}$ for which $T_1 = T_2$, that is

$$\frac{r_1}{r_2} = \frac{||\mathbf{V}_i(0)||}{||\mathbf{V}_2(0)||}\frac{\eta_2}{\sin\eta_2}\frac{\sin\eta_1}{\eta_1}\,. \tag{7.5}$$

This equation together with the equation

$$\begin{aligned} \mathbf{r}_1(0) + r_1\left[\frac{\mathbf{V}_1(0)}{||\mathbf{V}_1(0)||}, \mathbf{V}_1(0)^+\right]\begin{bmatrix}\cos\eta_1\\ \sin\eta_1\cos\xi_1\\ \sin\eta_1\sin\xi_1\end{bmatrix} &= \mathbf{r}_2(0)\\ +r_2\left[\frac{\mathbf{V}_2(0)}{||\mathbf{V}_2(0)||}, \mathbf{V}_2(0)^+\right]\begin{bmatrix}\cos\eta_2\\ \sin\eta_2\cos\xi_2\\ \sin\eta_2\sin\xi_2\end{bmatrix} & \end{aligned} \tag{7.6}$$

are four scalar equations for determining the six variables $r_i, \eta_i, \xi_i$; the solution is therefore a 2-dimensional set

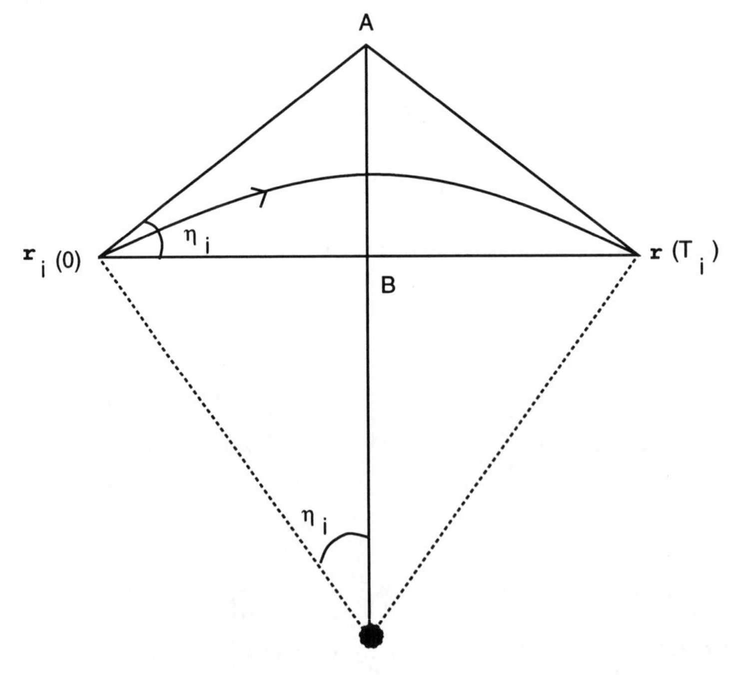


FIGURE 7.3.

## 7.3 Results

The following approach is used in [1] for the mini-max strategy. The pursuer tries to minimize the miss distance, while the evader tries to maximize some cost functional $J$. Given $\mathbf{u}_1$, the pursuer chooses $\mathbf{u}_2^* = \mathbf{u}_2^*(\mathbf{u}_1)$ and time $T^* = T^*(\mathbf{u}_1)$ such that

$$\min_{\substack{0 \le t \le T_{\max} \\ \mathbf{u}_2 \in U_2}} \|\mathbf{r}_2(t) - \mathbf{r}_1(t)\|$$

is achieved at $\mathbf{u}_2 = \mathbf{u}_2^*$ , $t = T^*$; here $T_{\max}$ is a prescribed large enough time. The evader is provided with this information, and he will choose $\mathbf{u}_1^*$ such that

$$\max_{\mathbf{u}_1 \in U_1} J(T^*(\mathbf{u}_1), \mathbf{u}_2^*(\mathbf{u}_1), \mathbf{r}_1(T^*(\mathbf{u}_1)), \mathbf{r}_2(T^*(\mathbf{u}_1))$$

is achieved at $\mathbf{u}_1 = \mathbf{u}_1^*$; here $J(t, \mathbf{u}_2, \mathbf{r}_1, \mathbf{r}_2)$ is given function. Finally, the pursuer will make the choice $\mathbf{u}_2^* = \mathbf{u}_2^*(\mathbf{u}_1^*)$.

This mini-max algorithm is updated at describe times $t_0 < t_1 < t_2 < \cdots$. The choice of $J$ in [1], for each update, is simply the time to interception. Thus, the evader considers all the intercept points $\mathbf{r}$ on the intercept surface, and then decides to choose the $\mathbf{u}_1$ for which the arrival time to $\mathbf{r}$ is maximal, i.e., the evader wishes to maximal his survival time if both vehicles were to travel with constant accelerations.

The numerical experiments described in [1] show that the convergence or divergence of the above algorithm is closely related to the structure of the intercept set. To explain this connection, we divide the 3D space into two sets:

$S_1$, where the evader gets to first, i.e., if $\mathbf{r} \in S_1$ then $T_1(\mathbf{r}) < T_2(\mathbf{r})$, and $S_2$ where the pursuer gets to first. The intercept surface is the boundary between the two sets, and it is not an algebraic surface. As shown in [1] $S_1$ consists of two parts: one part which contains the point $\mathbf{r}_1(0)$ and is contained in an ellipsoid, and another part which is contained in a cone whose axis is $\mathbf{r}_2(0) + \lambda \mathbf{V}_2(0)$ for $-\infty < \lambda < -\lambda_0$ (for some $\lambda_0 > 0$); it is always assumed that $\|\mathbf{V}_2(0)\| > \|\mathbf{V}_1(0)\|$. The ellipsoid and the cone are shown dotted in Figure 7.4. The intercept surface lies in the boundary of the dotted region.

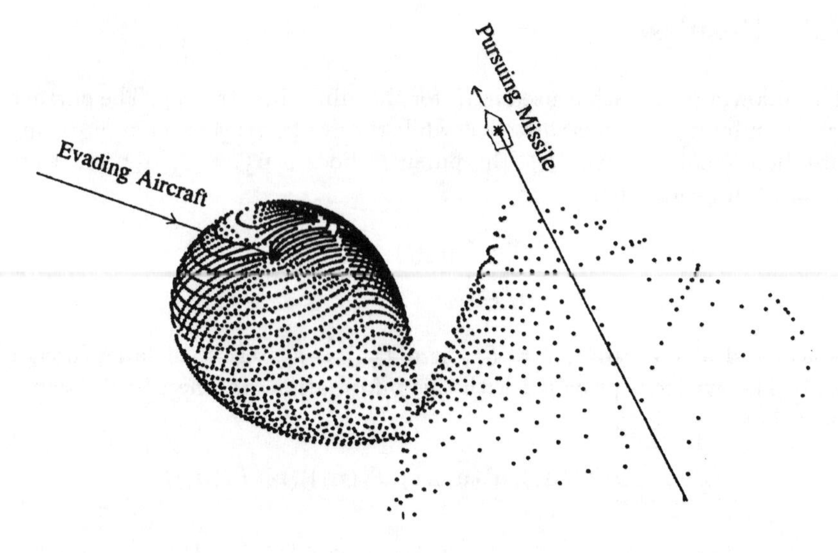


FIGURE 7.4.

In some cases the two components of $S_1$ are connected and in other cases they are disconnected. This depends on two quantities:

(i) the ratio $q = \dfrac{||\mathbf{V}_1(0)||}{||\mathbf{V}_2(0)||}$ , and

(ii) line-of-sight (los) angle which is the angle between $\mathbf{V}_2(0)$ and $\mathbf{r}_1(0) - \mathbf{r}_2(0)$ (see Figure 7.2); we set $s =$ los.

Figure 7.5 describes a curve $q = H(s)$ established in [1] such that for any $(q, s)$ below the curve $S_1$ has two components and for some $(q, s)$ above the curve there are examples where $S_1$ is connected.

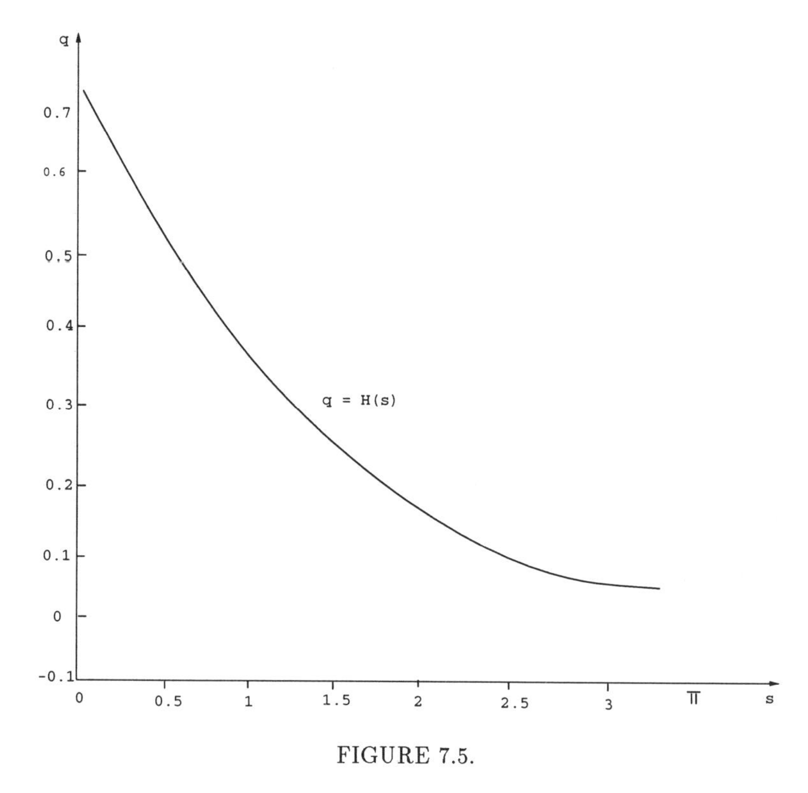


FIGURE 7.5.

Anywhere below the curve aircraft cannot escape, i.e., the mini-max iterative algorithm converges. Anyway above the curve examples are found where airplane can escape and the iterative algorithm defined in [1] for solving (7.5), (7.6) diverges.

*Problem (1).* Can this be proved rigorously?

## 7.4 Differential games approach

The above approach allows both vehicles to maneuver in 3D space and yet it is fast enough to run real-time on today's flight computers. Furthermore, the connection between the constraints and the intercept surface can help the pilot make a missile launch decision.

The present approach however does not take gravity into account, nor does it allow for nonlinear dynamics and nonlinear constraints; furthermore, it restricts the control variables $\mathbf{u}_1, \mathbf{u}_2$ to piecewise constant variables. On the other hand the approach of differential games [2] does not impose any of the above restrictions, although it is much more computationally intensive. To describe the differential game approach it is convenient to lump the variables $\mathbf{r}_1, \mathbf{V}_1, \mathbf{r}_2, \mathbf{V}_2$ together as a 12-dimensional variable $x$, and to denote the control variables by $y$ (for the evader) and $z$ (for the pursuer). The dynamical system is

$$\begin{aligned} \frac{dx}{dt} &= f(t,x,y,z) , \\ x(t_0) &= x_0 , \end{aligned} \tag{7.7}$$

and the payoff is

$$J(y,z) = g(\tilde{t}, x(\tilde{t})) + \int_{t_0}^{\tilde{t}} h(t,x(t),y(t),z(t))dt \tag{7.8}$$

where $\tilde{t}$ is the time when the game ends. Player $y$ chooses a "strategy" $y = y(t,x)$ and player $z$ chooses a strategy $z = z(t,x)$. Upon insertion into (7.7) we get a trajectory $x(t)$ and we can then compute the payoff $J$, which we shall also write as $J(y(t,x),\ z(t,x))$.

Player $y$ wants to maximize $J$ and player $z$ wants to minimize it. If $y^*(t,x)$ and $z^*(t,x)$ are such that

$$J(y^*(t,x),z) \geq J(y^*(t,x),z^*(t,x)) \geq J(y,z^*(t,x))$$

then $(y^*,z^*)$ is called a *saddle point*, and

$$V(t_0,x_0) \equiv J(y^*,z^*)$$

is called the *value* of the game. Similarly one defines the value $V(t,x)$ for any $(t,x)$.

The function

$$\tilde{H}(t,x,y,z,p) = h(t,x,y,z) + p \cdot f(t,x,y,z)$$

is called the *Hamiltonian*. Suppose

$$\max_{y \in Y} \min_{z \in Z} \tilde{H}(t,x,y,z,p) = \min_{z \in Z} \max_{y \in Y} \tilde{H}(t,x,y,z,p) \tag{7.9}$$

where $Y$ and $Z$ are sets of admissible controls for $y$ and $z$, respectively, and let $y^0(t,x,p)$ and $z^0(t,x,p)$ be functions satisfying:

$$\max_{y \in Y} \tilde{H}(t,x,y,z,p) = \tilde{H}(t,x,y^0(t,x,p),z,p) ,$$

$$\min_{z \in Z} \tilde{H}(t,x,y,z,p) = \tilde{H}(t,x,y,z^0(t,x,p),p) .$$

Set

$$H(t,x,p) = \tilde{H}(t,x,y^0(t,x,p),z^0(t,x,p),p) .$$

Then the value $V$ is a solution to the Hamilton–Jacobi–Bellman (HJB) equation

$$\frac{\partial V}{\partial t} + H(t, x, \nabla V) = 0 \tag{7.10}$$

subject to terminal conditions which, in case (7.8) with $\tilde{t} = \text{const.} = T$, reads:

$$V(T, x) = g(T, x) \; . \tag{7.11}$$

If $\tilde{t}$ is the intercept time (where $\mathbf{r}_1(t) = \mathbf{r}_2(t)$), then (7.11) is replaced by

$$V = g \quad \text{on the set } (\tilde{t}, x(\tilde{t})) \text{ where } \mathbf{r}_1(\tilde{t}) = \mathbf{r}_2(\tilde{t}) \; . \tag{7.12}$$

The solution to (7.10) (7.11) is typically not continuously differentiable, only Lipschitz continuous. On the other hand there are generally infinitely many Lipschitz continuous (or even piecewise continuously differentiable) solutions. A solution to (7.10), (7.11) can be obtained by first adding a "viscosity" term $\varepsilon \nabla^2 V$ ($\varepsilon > 0$) to the left-hand side of (7.10); the new system becomes a parabolic PDE and it has a unique solution $V_\varepsilon$. It can then be shown [4][5] that a subsequence of the $V_\varepsilon$'s converges to a Lipschitz solution $V$ of (7.10), (7.11). It turns out that there is only one limit $V$; furthermore, (by [6] [7]) this solution can be characterized directly from the equation (7.10), i.e., without going to the parabolic auxiliary problem; it is called a *viscosity solution*:

The viscosity solution satisfies (7.10) in the sense that for any continuously differentiable function $\varphi$ the following holds:

(i) if $V - \varphi$ attains a local maximum at a point $(\tilde{x}, \tilde{t})$ then

$$\varphi_t(\tilde{x}, \tilde{t}) + H(\tilde{t}, \tilde{x}, \nabla\varphi(\tilde{t}, \tilde{x})) \geq 0 \; ;$$

(ii) if $V - \varphi$ attains a local minimum at a point $(\tilde{x}, \tilde{t})$ then

$$\varphi_t(\tilde{x}, \tilde{t}) + H(\tilde{t}, \tilde{x}, \nabla\varphi(\tilde{t}, \tilde{x})) \leq 0 \; .$$

It can be proved that a continuously differentiable solution of (7.10) is a viscosity solution, and that a viscosity solution satisfies (7.10) at each point where it is continuously differentiable [6][7].

To compute the viscosity solution one cannot use standard finite difference schemes; the "viscosity solution" character must somehow be preserved in the approximation scheme. Recent progress on such numerical approaches is described in [8] and the references therein. For more theoretical numerical methods see [9][10][11][12].

*Problem (2).* Consider the dynamical system (7.1) with variables $\mathbf{u}_i$ which vary continuously in time. Write down the HJB equation.

It would be interesting to compute the solution of this HJB equation subject to the boundary condition (7.12) with cost functional

$$g(t,x) = t \ , \ h(t,x,y,z) \equiv 0 \ ,$$

and to compare the results with those in [1]. This will give an indication as to how close the mini-max solution is to the differential game solution.

## 7.5 References


[1] M. Elgersma and B. Morton, *Progress in estimation and control for air-launched missiles: Part I, A real-time 3D mini-max pursuit-evader algorithm*, Honeywell Technology Center, Minneapolis, December 1995.

[2] R. Isaac, *Differential Games*, Wiley, New York (1965).

[3] A. Friedman, *Differential Games*, Wiley–Interscience, Wiley, New York (1971).

[4] A. Friedman, *Differential Games*, Regional Conference Series in Mathematics, No. 18, Amer. Math. Soc., Providence, R.I. (1994).

[5] A. Friedman, *Differential Games*, in Handbook of Game Theory, Vol. 2, editors R.J. Aumann and S. Hart, pp. 781–799, Elsevier, Amsterdam (1994).

[6] M.G. Crandall and P.L. Lions, *Viscosity solutions of Hamilton–Jacobi equations*, Trans. Amer. Math. Soc., 277 (1983), 1–42.

[7] M.G. Crandall, L.C. Evans and P.L. Lions *Some properties of viscosity solutions of Hamilton–Jacobi equations*, Trans. Amer. Math. Soc. 282 (1984), 487–502.

[8] D. Adalsteinsson and J.A. Sethian, *A level set approach to a unified model for etching, deposition, and lithography I: Algorithms and two-dimensional simulations*, J. Computational Physics, 120 (1995), 128–144.

[9] M.G. Crandall and P.L. Lions, *Two approximations of solutions of Hamilton–Jacobi equtions*, Mathematics of Computations, 43 (1984), 1–19.

[10] P.E. Souganidis, *Approximation schemes for viscosity solutions of Hamilton-Jacobi equations*, J. Diff. Eqs., 59 (1985), 1–43.

[11] G. Barks and P.E. Souganidis, *Convergence of approximation schemes for fully nonlinear second order equations*, Asymptotic Analysis, 4 (1991), 271–283.

[12] P.L. Lions and P.E. Souganidis, *Convergence of MUSCL and filtered schemes for scalar conservation laws and Hamilton–Jacobi equation*, Numer. Math., 69 (1995), 441–470.

# 8

# Formation of photographic images

The development of latent image sites in photographic film involves diffusion of electrons in silver halide crystals with trapping sites. Electron $e^-$ captured at a trapping site can either escape from the trap or else attract an interstitial silver ion $Ag^+$ to form neutral silver atom $Ag^0$ at the trapping site. The process can be repeated with other electrons. When at least four neutral silver atoms are formed at a trap, latent image is assured at this site. Monte Carlo simulation is used to compute the motion of the free electrons as well as the process that takes place at the traps. On January 18, 1996 Peter Castro from Eastman Kodak presented the Gurney–Mott–Hamilton–Bayer theory which explains in more detail how the latent image is formed; this was also described in [1, Chap. 14]. He then outlined a new approach which includes the effect of the electric field, and simulation which is based on non-standard stochastic structure. He presented some open mathematical problems whose solution could provide better Monte Carlo results.

## 8.1 The Gurney–Mott–Hamilton–Bayer theory

A photographic film consists of several emulsion layers and gelatine separator layers. An emulsion layer contains silver halide grains of 0.5 to 2 $\mu m$ in diameter. When a picture is taken, photons of incident light enter the film layers and are scattered and absorbed by silver halide grains. An absorbed photon releases a mobile electron $e^-$ and a positive hole. The electrons diffuse to trapping sites either within the grain or on its surface (at "kink sites"). The silver halide grain is a crystal made up of chlorine, bromine or iodine atoms, and silver atoms. There are also interstitial silver ions $Ag^+$ in the crystal, and they are attracted to the negative charge in the trap site, forming with $e^-$ a neutral silver atom $Ag^0$. If not too long a time passed, a trap site is capable of trapping a second electron, after which the process repeats itself and another neutral silver atom is created at the site. However, a silver atom may also decompose into a mobile electron $e^-$ which diffuses back into the crystal and an interstitial silver atom $Ag^+$. Four neutral silver atoms in a trapping site form a stable configuration, and this site is a latent image site. When the film is developed latent image sites participate in image forming dye reactions.

The above theory is due to Gurney and Mott ([2]; see also [3, Chap. 12]), and was further developed by Hamilton and Bayer [4].

Figure 8.1 shows the motion of an electron $e^-$ and hole $Br^+$ in bromide. Electron mobility is $\sim 5 \times 10^8 \mu m^2$ sec$^{-1}$ whereas the hole mobility is $\sim 5 \times 10^6 \mu m^2$ sec$^{-1}$, so it is much smaller. Film exposure is Poisson distributed, measured by $E$ photons/area/time for $D$ time units;

$$D \sim 10^{-4} - 10^4 \text{ sec},$$

$$E \sim 2 \times 10^{-4} - 1024 \times 10^4 \text{ sec}^{-1} \text{ grain}^{-1}.$$

The electron trap radius is $\sim 0.005 \mu m$, and the number of electron traps in one grain is $\sim 1500$. (The traps in Figure 8.1 are on the grain's boundary.)

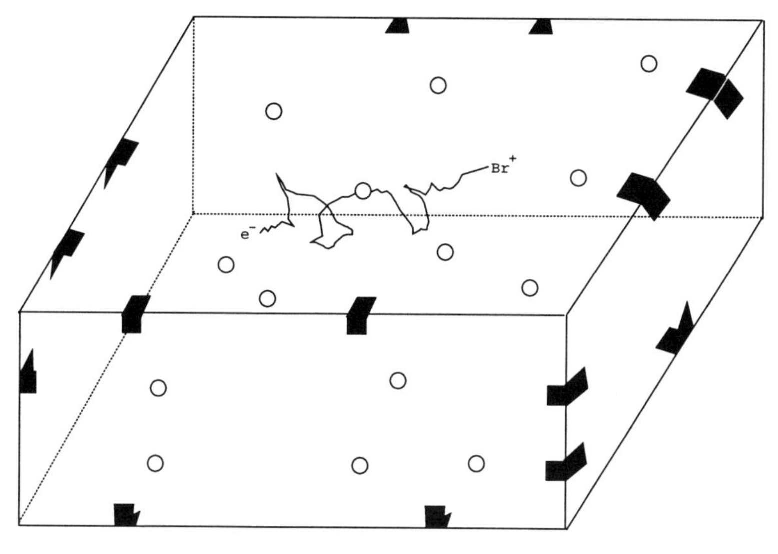


FIGURE 8.1.

The holding time of electron in trap in nucleation phase is $\sim 10 - 100 ns$ ($ns = 10^{-9}$ sec.). Time for $Ag^+$ to neutralize trapped $e^-$ is $\sim 0.1\mu$ sec. $Ag^0$ decay time is $\sim 0.1$ sec.

The basic assumptions of the Gurney–Moth–Hamilton–Bayer theory are:

(1) Charge carriers — electrons and holes — move about the grain volume in a manner that does not depend upon position; that is, there is a constant mobility, or carrier speed, for each carrier type which is valid for that carrier type everywhere in the grain.

(2) 'Traps' are present in the volume and at the surface of each grain and carriers become immobilized at traps according to the trapping cross sections, density of traps, and density of carriers (number per unit volume).

(3) Holding times of carriers in traps are exponentially distributed with mean holding time dependent on trap configuration.

(4) Certain trapping and detrapping events are so fast in comparison with other events that electrons and holes can be assumed to be in equilibrium among several states.

(5) Random times between events in a grain are exponentially distributed.

Figure 8.2 is a basic diagram which describes the simulations. The absorption part is very fast compared with the nucleation and growth, and the simulations therefore assume equilibrium states for these fast processes (as alluded to in (4)). Note that holes may be lost by migrating to the surfaces of the grain

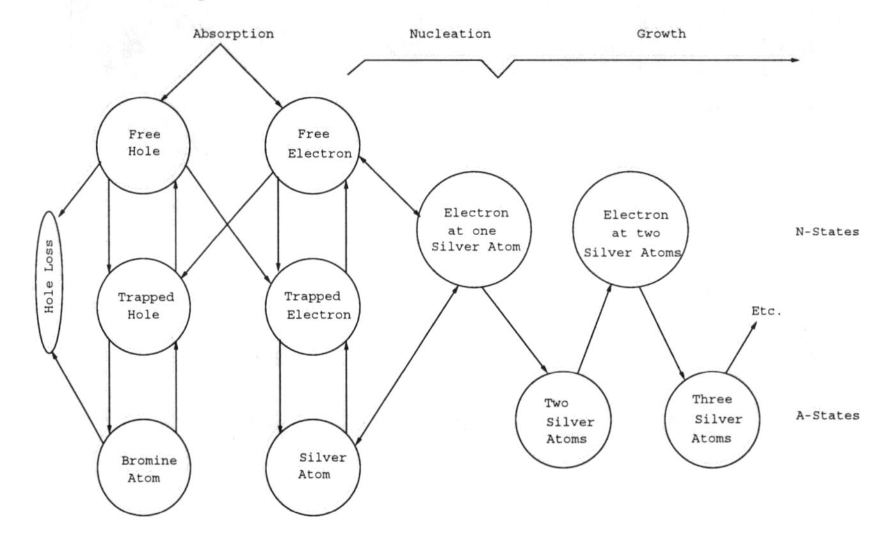


FIGURE 8.2.

In the precipitation of silver halide grains several physical quantities can be controlled: Number of electron traps, electron trap radius, holding time of electron in trap in nucleation phase, and position of traps in grain. For this reason, simulating the nucleation and growth processes is important for improving the quality of film.

## 8.2 Monte Carlo simulation

The Monte Carlo simulation for absorption of free electrons and for nucleation and growth was described in [1, Chap. 14] when the traps are distributed on the surface of the grain. Figure 8.3 shows simulation results (taken over 500 grains) for four different exposure durations, ranging from 0.0001 sec. to 0.1 sec. For each exposure duration there are 6 curves; the first curve (from left to right) shows the fraction of grains for which there is at least one trap with two silver atoms $Ag^0$; the second curve shows the fraction of grains for which there is at least one trap with three silver atoms, the third curve corresponds to four silver atoms in at least one trap, etc. The simulation results (marked by "+") are qualitatively compatible with experiments.

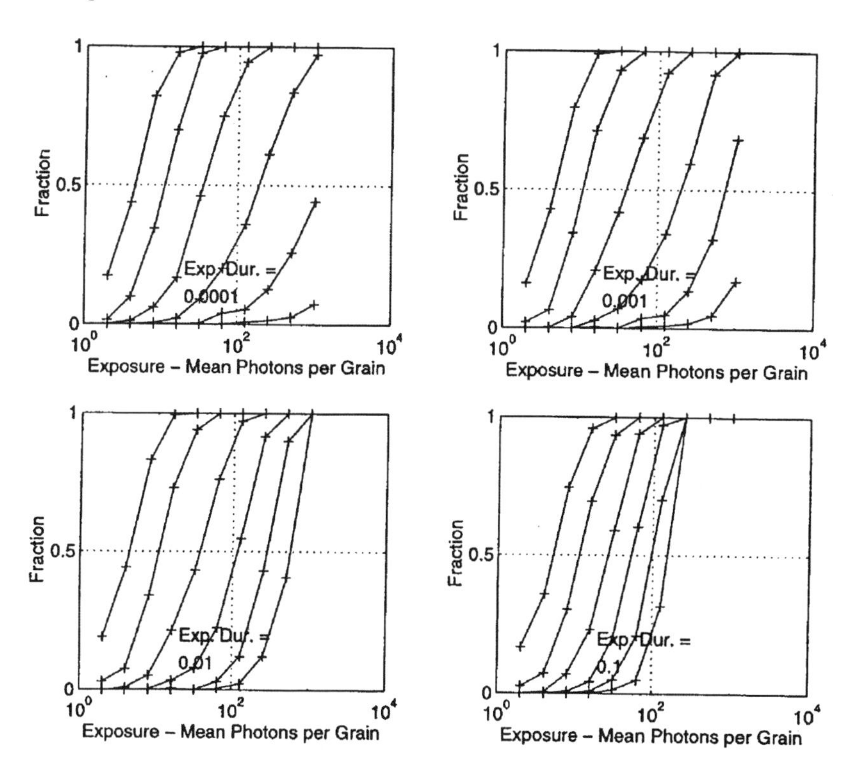


FIGURE 8.3.

Even though the simulation is very time consuming, the model is not flexible enough; for example, it does not take into account properties which depend on position in the grain. Of greater concern is the fact that the model does not consider the effect of the electric field on the electron motion. There are crystal defects on the grain surface leading to a negative surface charge. Mobile positive carriers respond to this, producing a positive space charge inside the grain.

In the above simulation the electron motion at the surface of a grain is subjected to partial absorption and partial reflection. If we use PDE as basis for the Monte Carlo simulation, then what happens at the boundary may be described by boundary conditions

$$\frac{\partial u}{\partial n} + Au = f \quad \text{or} \quad u = g \quad (A \ \text{constant})$$

at different portions of the boundary. (Here $u$ satisfies the heat equation outside the traps.) Since the boundary is divided into many small portions of these two types, it may be useful to "average" or "homogenize" the above oscillating boundary conditions. Work in this direction can be found in [5][6] and the references therein.

## 8.3 New model for electron motion

The new model includes the electric field $\mathbf{E}$ and is based on the following assumptions:

(i) Electron travels in classical trajectories between collisions with phonons.

(ii) Electric field is position dependent, and Newton's law holds:

$$\ddot{\mathbf{x}} = \frac{q}{m_e}\,\mathbf{E}(\mathbf{x}) \ , \ \mathbf{x}(t_k) = \mathbf{x}_k \ , \ \dot{\mathbf{x}}(t_k) = \mathbf{v}_k \tag{8.1}$$

where $q$ is the electron charge and $m_e$ is the electron mass.

(iii) Each collision erases previous momentum, i.e., the $\mathbf{v}_k$ are independent identically distributed (i.i.d) variables.

(iv) The time $t$ between collisions is exponentially distributed; this renders the process $\mathbf{x}(t)$ Markovian.

(v) Each trajectory from $\mathbf{x}(t_k)$ to $\mathbf{x}(t_{k+1})$ is short compared to change in field strength, so that we can take

$$\mathbf{x}_{k+1} = \mathbf{x}_k + \mathbf{v}_k(t_{k+1} - t_k) + \frac{q}{2m_e}\,\mathbf{E}(\mathbf{x}_k)(t_{k+1} - t_k)^2 \ . \tag{8.2}$$

The objective is to get an efficient simulation of the $\{\mathbf{x}_k\}$ process with $\mathbf{E}(\mathbf{x})$ determined by some charge density of the grain surface and mobile interstitial carriers (space charge) of the opposite sign which come into equilibrium with the resulting field.

The process (8.2) was simulated with uniform time steps $\Delta t = t_{k+1} - t_k$. However Castro would like to simulate the process under the more realistic assumption (iv) of exponentially distributed time between collisions. In this case note that if $\mathbf{v} = (v_1, v_2, v_3)$ is a velocity where each component $v_i$ is Gaussian with mean zero and variance $\sigma^2$, i.e., its distribution function is

$$f_{v_i}(u) = \frac{1}{\sqrt{2\pi}\,\sigma}\, e^{-u^2/(2\sigma^2)} \ ,$$

then the distribution of the speed $v = (v_1^2 + v_2^2 + v_3^2)^{1/2}$ is Maxwellian with density

$$f_v(u) = \frac{2}{\sqrt{2\pi}} u^2 e^{-u^2/(2\sigma^2)} .$$

Next compute the distribution function $f_r(x)$ of the distance $r$ between collisions, assuming that the time $t$ between collisions has density function

$$f_t(s) = \frac{1}{\tau} e^{\;s/\tau}$$

where $\tau$ is the mean intercollision time. Using the fact that $r = vt$, one finds that

$$f_r(x) = \frac{2}{\sqrt{2\pi}\ \sigma\tau} \int_0^\infty \exp\left[-s - \frac{x}{\sqrt{2\sigma\tau}\ \sqrt{s}}\right] ds .$$

The mean and variance of the step length are then

$$\mu_r = \frac{4\sigma\tau}{\sqrt{2\pi}} , \quad \sigma_r^2 = (6 - 8/\pi)\sigma^2\tau^2 .$$

The simulation under the assumption (iv) is more precise than the simulation under the assumption of uniform time steps $\Delta t$, but it requires (as indicated by the density $f_r(x)$) short time and spatial steps and therefore longer simulation time.

To overcome this difficulty Castro is looking more carefully at the process (8.2). Since $\delta\mathbf{x}_{k+1} = \mathbf{x}_{k+1} - \mathbf{x}_k$ is independent of $\mathbf{v}_k$ and $\mathbf{v}_k$ has mean zero and is independent of $\delta t_{k+1} = t_{k+1} - t_k$, by taking the conditional expectation of (8.2) with respect to $\mathbf{x}_k$ we get

$$E[\delta\mathbf{x}_{k+1}|\mathbf{x}_k] = \frac{q}{2m_e} \mathbf{E}(\mathbf{x}_k)2\tau . \tag{8.3}$$

Similarly,

$$E[\text{Var}\ (\delta\mathbf{x}_{k+1})|\mathbf{x}_k] = \text{Var}\ (\mathbf{v})2\tau + \left[\frac{q}{2m_e} \mathbf{E}(\mathbf{x}_k)\right]^2 20\tau . \tag{8.4}$$

*Problem (1).* Can the process (8.2) be approximated, as $\tau \to 0$, by a diffusion process?

Clearly if $(t_{k+1}-t_k)^2$ is replaced by $t_{k+1}-t_k$ in the last term in (8.2) and if $t_{k+1}-t_k$ is further taken to be non-random, then, as $\max_k(t_{k+1}-t_k) \to 0$, the process converges to a diffusion process with drift term $(q/2m_e)\mathbf{E}(\mathbf{x})$ and a diffusion term which depends on $\mathbf{v}$. Castro suggested that, under the present assumption of exponentially distributed $\delta t_k$, the invariance principle of Stroock and Varadhan [7] may solve Problem (1). This principle asserts that, under some general conditions, a sequence of discrete Markov processes converges to a diffusion process.

The second question raised by Castro is the following: The mobile, positive carriers which respond to the field and modify it are themselves shallow electron traps. If we add them to the problem as potential wells, can we get a "hopping model" for electron jumps between traps? This would bypass all uninteresting diffusion-type motion between trapping events.

An example where a hopping model provides a good answer arises in the "exit" problem for the stochastic differential system

$$d\xi = b(\xi)dt + \varepsilon\sigma(\xi)dw \tag{8.5}$$

in an $n$-dimensional domain $D$, where $\varepsilon$ is a small parameter. Assume that the dynamical system

$$dx = b(x)dt$$

has a finite number of "traps" in $D$, that is, points of stable equilibrium. In computing quantities such as the principal eigenvalue of the elliptic operator corresponding to (8.5), the leading term is determined just by the finite Markov chain with states which are identified with the traps and with transition probabilities that can be calculated from the data $b(x), \sigma(x)$; see [8].

## 8.4 References


[1] A. Friedman, *Mathematics in Industrial Problems, Part 2,* IMA Volume 24, Springer-Verlag, New York (1989).

[2] R.W. Gurney and N.F. Mott, *Electronic Processes in Ionic Crystals,* Oxford (1938).

[3] F.C. Brown, *The Physics of Solids,* Benjamin New York (1967).

[4] B.E. Bayer and J.F. Hamilton, *Computer investigation of a latent image model,* Journal of the Optical Society of America, 55 (4) (1965), 439–452.

[5] M.J. Ward and J.B. Keller, *Strong localized perturbation of eigenvalue problems,* SIAM J. Appl. Math., 53 (1993), 770–798.

[6] A. Friedman, C. Huang and J. Yong, *Effective permeability of the boundary of a domain,* Comm. PDE, 20 (1995), 59–102.

[7] D.W. Stroock and S.R.S. Varadhan, *Diffusion processes with continuous coefficients, II,* Comm. Pure Appl. Math., 22 (1967), 479–530.

[8] A.D. Ventcel, *On the asymptotic behavior of the greatest eigenvalue of a second order elliptic differential operator with a small parameter in the higher derivatives,* Soviet Math. Doklady, 13 (1972), 13–17 [Dokl. Akad. Nauk SSSR, 202 (1972), no. 1].

# 9

# Convective-diffusive lattice-gas models for flow under shear

Coating flow is fluid flow that is used in many industrial processes for covering a surface area. For example, a curtain coating process is used in the manufacturing of photographic films. In curtain coating, a liquid film, formed on an inclined slide (as shown in Figure 9.1, falls under the gravity onto a substrate (film base) which is moving with uniform velocity $\vec{U}$.

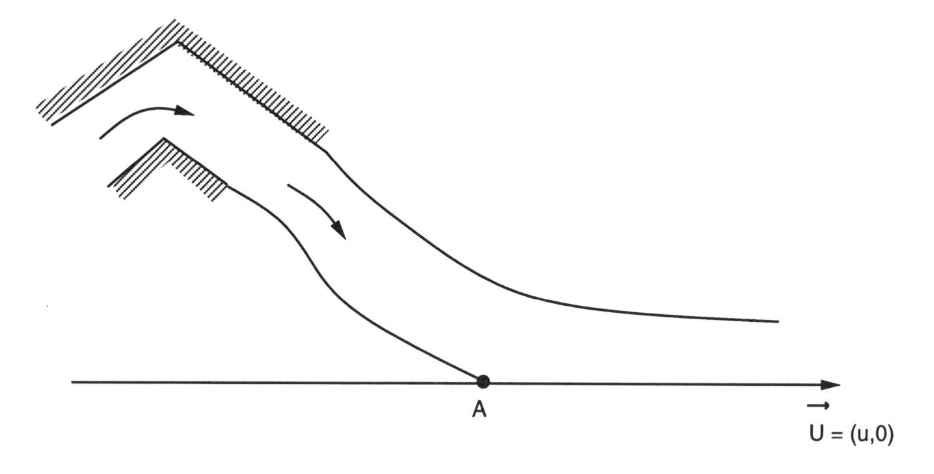


FIGURE 9.1.

There are other coating methods, such as dip coating, extrusion coating, roll coating, etc.; we refer to [1, Chap. 3] for references.

On February 9, 1996 Yitzhak Shnidman from Kodak Research Labs has discussed various models for coating flows and, more generally, for forced wetting. The most important question is what happens near the contact line $A$; we assume 3-d flow independent of the $y$-direction, so that the analysis is basically 2-d in the $(x, z)$-plane, and $A$ is actually a point. After explaining the difficulties encountered by the purely hydrodynamic theories, he proceeded to develop a lattice-gas model which includes both convection and diffusion. His model, which is based on the mean field assumption, extends also to 2-phase flows with spinodal decomposition. This work is, in part, jointly with Adil Khan from the University of Rochester; an earlier version was developed in [2]. The motivation for this work is to improve the understanding and modeling of coating processes and, in particular, to develop better prediction of the behavior of the flow near the contact point as a function of the chemical parameters of the fluids and the material constants of the substrate.

## 9.1 The hydrodynamic model

*Dynamic wetting* occurs when one fluid displaces another at the substrate. We distinguish between *spontaneous spreading* and *forced wetting*. Spontaneous spreading is a time-dependent relaxation to equilibrium, such as the spreading of a drop on a plane. In forced wetting, the three interfaces solid/fluid 1, solid/fluid 2, and fluid 1/fluid 2 meet at the triple (contact) line $A$, and the interface between the two fluids at $A$ is driven at steady dynamic angle $\theta = \theta(u)$, where $u$ is the relative velocity between $A$ and the solid. Figures 9.1 and 9.2 show typical realizations of forced wetting. Figure 9.3(a) is a Poiseiulle flow (pressure gradient from the left pushes the flow); in Figures 9.3(b) and (c) the "solid face" is a plate and a tape, respectively.

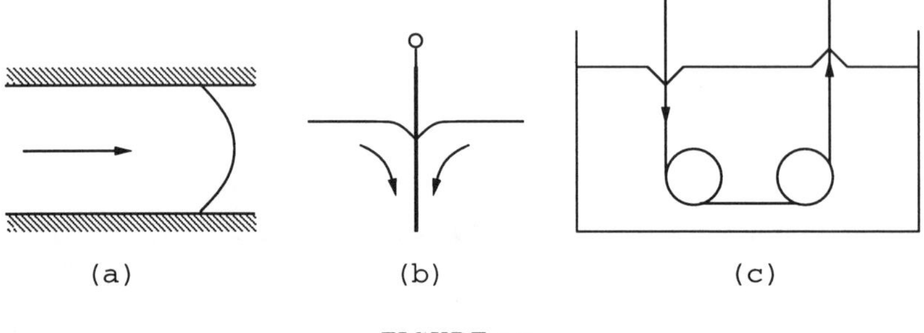


FIGURE 9.2.

Figure 9.3 describes the generic situation near the contact line. The dynamic angle at $\theta(u)$ reduces to the static angle $\theta_0$ when $u = 0$. In the vicinity of the contact line $A$ the interfaces are actually not linear, but rather "near" linear.

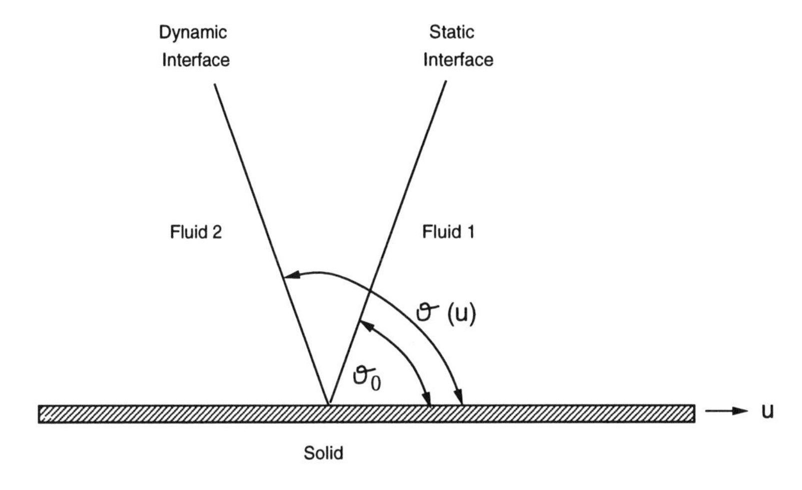


FIGURE 9.3.

Since industry would like to move the product faster, it is important to study the behavior of $\theta(u)$ for $u$ large.

The conventional hydrodynamic approach is based on the incompressible Navier–Stokes equations in the bulk with no-slip boundary conditions, assuming "sharp" (infinitesimally thin) interfaces:

$$\mathrm{Re}\left[\frac{\partial \vec{u}}{\partial t} + \vec{u} \cdot \nabla \vec{u}\right] = \frac{1}{\rho}\nabla T + \vec{F} \qquad \text{(conservation of momentum)}, \tag{9.1}$$

$$\nabla \cdot \vec{u} = 0 \qquad \text{(incompressibility)} \tag{9.2}$$

in each liquid, where $\vec{u} = (u_1, u_2, u_3)$ is the fluid velocity, $T$ is the stress tensor

$$T_{ij} = -\delta_{ij} p + \left(\frac{\partial u_i}{\partial x_j} + \frac{\partial u_j}{\partial x_i}\right),$$

Re= Reynolds number,

$\rho =$ fluid density (= constant),

$p =$ pressure

and $\vec{F}$ is force (usually gravity). At the fluid/solid interface

$$\vec{u} = \text{ velocity of the solid (no-slip)} \tag{9.3}$$

and, at the fluid/fluid interface,

$$\vec{\tau} \cdot T\, \vec{n} = 0 \qquad \text{(no shear)}, \tag{9.4}$$

$$[\vec{n} \cdot T\, \vec{n}] = \frac{1}{Ca}\, K \tag{9.5}$$

where $Ca$ is the capillary number and $K$ the curvature; here $\vec{n}$ is the normal and $\vec{\tau}$ is the tangent, and $[\ldots]$ in (9.5) means the jump across the interface.

The conditions (9.4), (9.5) together with the no-slip condition (9.3) near the contact point $A$ imply that the veelocity is not continuous at $A$; further, in any neighborhood $D_0$ of $A$ the energy dissipation rate is infinite, that is,

$$\int_{D_0} (\partial_i u_j + \partial_j u_i)^2 = \infty \; .$$

Since these conclusions are physically unacceptable, attempts have been made to modify the no-slip condition in some neighborhood $D$ of $A$. However, the choice of such a neighborhood $D$ is rather arbitrary, and there are ambiguities in the choice of the boundary conditions near $A$; we refer to [3],[4],[5],[6] and the references therein for various models that have been suggested. So far experimental evidence has not distinguished between these theories; experiments only probed distances larger than $1\mu m$ from the contact line, where all the theories predict approximately the same flow.

In a more recent work [7] Friedman and Velázquez consider the 2-d coating problem described in Figure 9.4 with one fluid only, i.e., the air is neglected.

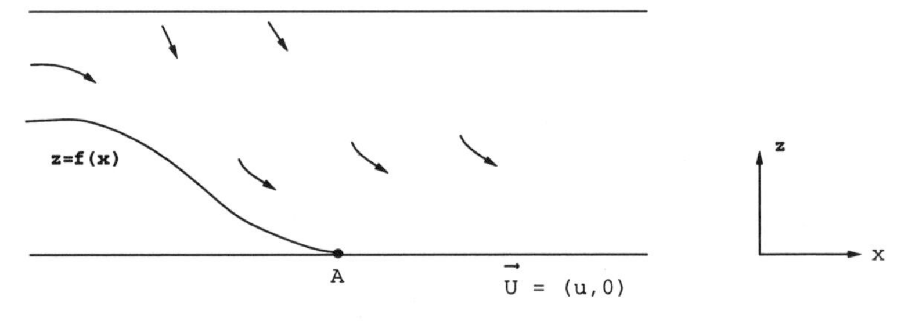


FIGURE 9.4.

They prove that the Navier–Stokes formulation with no-slip condition has a unique solution with $f' = 0$ at $A$. This means that the apparent dynamic angle will have to be computed as some average, not "too close" to $A$.

An entirely new approach was recently developed by Shikhmurzaev [8][9]. He introduces "surface phase" and quantities such as surface density, surface velocity and surface stress, and then applies conservation laws to derive a system of differential equations for the surface quantities, coupled to the equations in the bulk of the two fluids. Several phenomenological constants enter into these equations. He established numerically results such as the "rolling" motion near the contact line, which was also derived by Dussan [3] using an entirely different approach.

## 9.2 Molecular dynamics models

Blake and Haynes [10] developed a model of fluid flow based on the concept of lattice gas: fluid particles are in cages and undergo local motion due to the motion of cages, but they also hop to neighboring vacant cages at longer time scale.

The hopping is based on the model of Eyring [11] and Frenkel [12]. In equilibrium, for a molecule in one cage to jump into an unoccupied neigh boring cage—it must overcome energy barrier $\varepsilon$; see Figure 9.5(a).

The mean rates of hopping are given by

$$j_0 = A \exp\left(-\frac{\varepsilon}{k_B T}\right)$$

where $k_B$ is the Boltzmann constant and $T$ is the absolute temperature; because of symmetry there is no net flow. In non-equilibrium steady state (NESS) under shear stress $\vec{E}$ of magnitude $E$, the energy barriers become biased as indicated in Figure 9.5(b). Consequently, the hopping rates are not symmetric, and the shear-biased rates lead to net flow (i.e., current)

$$\begin{aligned} j &= A\left[\exp\left(-\frac{\varepsilon - \frac{1}{2}E\alpha\lambda}{k_B T}\right) - \exp\left(-\frac{\varepsilon + \frac{1}{2}E\alpha\lambda}{k_B T}\right)\right] \\ &= 2A \exp\left(-\frac{\varepsilon}{k_B T}\right)\sinh\left(\frac{E\alpha\lambda}{2k_B T}\right) \end{aligned} \tag{9.6}$$

where $\lambda$ is the spacing between cages and $\alpha$ is the area per site.

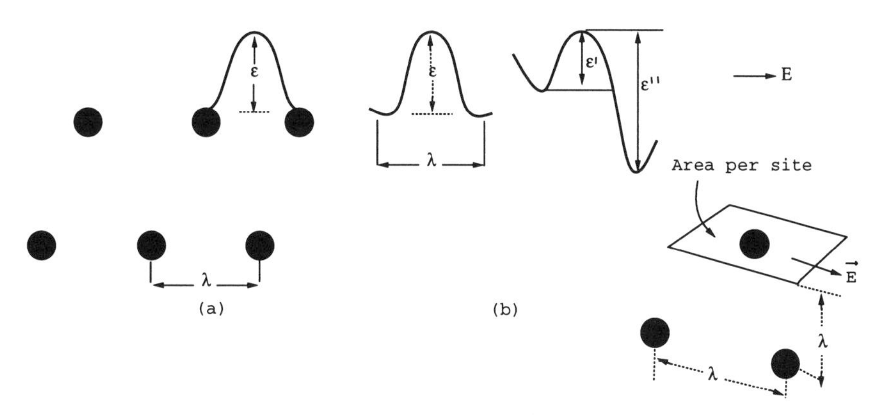


FIGURE 9.5.

Blake and Haynes argue as follows: In the static case the balance of tensions at the contact line is given by the well known Young equation

$$\gamma_{12} \cos\theta_0 = \gamma_{s1} - \gamma_{s2}$$

where $\gamma_{sj}$ is the surface tension at the solid/fluid 1 interface and $\gamma_{12}$ is the surface tension at the fluid 1/fluid 2 interface. In the dynamic case, the dynamic angle $\theta$ is a consequence of unbalanced tension $w$, and

$$w = \gamma_{12}(\cos\theta_0 - \cos\theta) \ . \tag{9.7}$$

This tension $w$ should lead, in the Eyring–Frenkel model, to hopping rates

$$K_{\pm} = K_0 \exp\left(\frac{\pm w}{2nk_BT}\right) \tag{9.8}$$

where $n = 1/(\alpha\lambda)$. Substituting $w$ from (9.7) in (9.8) and using the net-flow formula (9.6), we get the flow velocity $u$:

$$u = 2K_0\lambda \sinh\left(\frac{\gamma_{12}(\cos\theta_0 - \cos\theta)}{2nk_BT}\right) \ , \tag{9.9}$$

Unfortunately, this elegantly simple formula relating $\theta$ to $u$ does not take into account some important parameters of the system, such as the molecular interactions reflecting the chemistry of the fluid. Moreover, the flow velocities were assumed to be of a purely diffusive origin. Such dynamics lacks local conservation of momentum, which is the basis of hydrodynamic approaches. Real molecules in a liquid should obey local conservation of momentrum, except at moving walls, where momentum transfer occurs. In between diffusive hops between cages, real molecules are not static, but are rather convected by the surrounding fluid.

Thompson and Robbins [13] carried out molecular-dynamics simulations of two immiscible fluids confined between two solid walls under shear stress. They solve Newton's equations assuming that the potential between atoms (of the fluid) separated by distance $r$ is

$$Ar^{-12} - Br^{-6} \quad \text{(the Lennard–Jones potential)}.$$

Their conclusions are consistent with the macroscopic theories: Navier–Stokes equations with no-slip are valid away from the contact zone; near the contact zone there is velocity slip and enhanced stresses, and the flow is tangential to the interface. They show that near the contact line the no-slip boundary condition breaks down at the atomic scale for the first two atomic spacing. Figure 9.6 shows, for the case of boundary condition periodic in the $x$-direction, typical steady-state snapshots obtained in [13] of a phase separated two-species system between solid walls at different wall velocities.

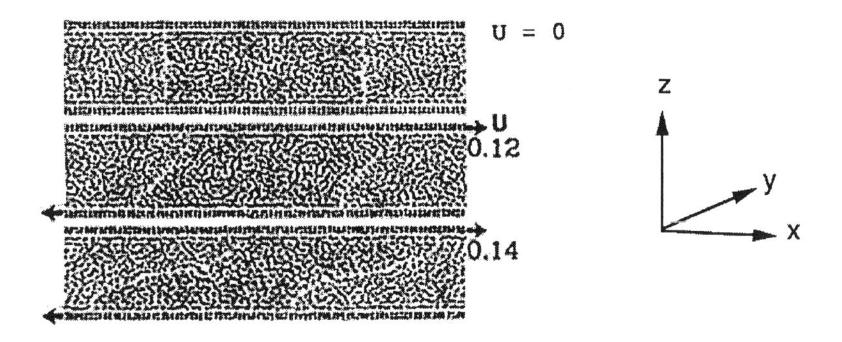


FIGURE 9.6.

## 9.3 Convective-diffusive lattice-gas model

We now describe the model of Shnidman which incorporates both convection and diffusion in the lattice-gas framework.

We deal with slab $\Lambda$ in which there are $N$ sites distributed on a cubic-lattice between two solid walls at $z = -a$ and $z = La$. Periodic boundary conditions are used along the $\widehat{x}$ direction. We will consider the case where the two walls are sheared with opposite steady velocities $\pm v_w$, like in Ref. [13]. Consider first the case of one molecular species. If a site $\vec{r}$ is occupied by a molecule then we write $s_{\vec{r}} = 1$; if the site is vacant then we write $s_{\vec{r}} = -1$. We take the Hamiltonian to be the one used in the Ising model,

$$H = -\sum_{\vec{r} \in \Lambda} \left[ \left( \sum_{k=1}^{6} \frac{1}{2} J s_{\vec{r}} s_{\vec{r} + \vec{e}_k} \right) + h_{\vec{r}} s_{\vec{r}} \right] \tag{9.10}$$

Here $J$ is a positive constant, corresponding to attractive ineractions between pairs of fluid molecules at adjacent sites, and $h_{\vec{r}}$ is the local chemical potential, which is used in expressing the interaction with the solid wall (cf. [13, Chap. 19]); these interactions are assumed to have a short range, i.e. $h_r$ vanishes everywhere except at the boundary layers; the $\vec{e}_k$ are the six unit vectors in the directions of the positive and negative $x$-$y$-$z$ axes. The *order parameter* is the statistical average of $s_{\vec{r}}$ taken over all configurations:

$$m_{\vec{r}} = \langle s_{\vec{r}} \rangle \ ;$$

it tells us in what state $\vec{r}$ is ($m_{\vec{r}} = 1$ means $\vec{r}$ is liquid, and $m_{\vec{r}} = -1$

means $\vec{r}$ is "dry"). The total density is defined by

$$n_{\vec{r}} = \frac{1}{2}\left(1 + m_{\vec{r}}\right) .$$

In the case of two molecular species, we define

$$S_{\vec{r}} = \begin{cases} 1 & \text{if } \vec{r} \text{ is occupied by species 1} \\ -1 & \text{if } \vec{r} \text{ is occupied by species 2} \\ 0 & \text{if } \vec{r} \text{ is vacant} \end{cases}$$

and take the Hamiltonian as in the Blume–Emery–Griffiths (BEG) model:

$$H = -\sum_{\vec{r} \in \Lambda} \left[ \frac{1}{2} \sum_{k=1}^{6} \left( J S_{\vec{r}} S_{\vec{r} + \vec{e}_k} + G S_{\vec{r}}^2 S_{\vec{r} + \vec{e}_k}^2 \right) + h_{\vec{r}} S_{\vec{r}} + \Delta_{\vec{r}} S_{\vec{r}}^2 \right] . \tag{9.11}$$

Here $G$ is another positive constant and $\Delta_{\vec{r}}$ is another chemical potential. We now introduce two order parameters:

$$\text{phase separation } m_{\vec{r}} = \langle S_{\vec{r}} \rangle, \text{ and}$$

$$\text{total density } n_{\vec{r}} = \langle S_{\vec{r}}^2 \rangle .$$

We are going to introduce convection and diffusion. Convection takes place between hops, whereby cages are convected by imposed velocity field which results from mechanical perturbation out of equilibrium. Diffusion, on the other hand, is due to molecules hop from their cages to neighboring vacant cages, and it is a consequence of dissipative relaxation toward equilibrium.

Denoting the site velocities by $\vec{v}_r$, we also add kinetic energy to the Hamiltonian:

$$K = \frac{1}{2} \sum_{\vec{r} \in \Lambda} \frac{1}{2}(1 + s_{\vec{r}}) \mu c^2 \, \vec{u}_{\vec{r}}^{\,2} \quad \text{(Ising)} , \text{ or}$$

$$K = \frac{1}{2} \sum_{\vec{r} \in \Lambda} \mu c^2 S_{\vec{r}}^2 u_{\vec{r}}^2 \quad \text{(BEG)}.$$

Here $\mu$ is the molecular mass which, for simplicity is assumed to be identical for both species, and $c$ has the dimensions of velocity, so that $\vec{u}$ is dimensionless. Henceforth we choose

$$\frac{\mu c^2}{k_B T} = 1 .$$

Next we make three assumptions:

(1) mean-field approximation,

(2) isothermal motion, and

(3) local equilibrium.

The mean-field approximation is

$$P(\{s_{\vec{r}}\ ;\ \vec{u}_{\vec{r}}\}) = \prod_{\vec{r}\in\Lambda} p_{\vec{r}}(s_{\vec{r}})q_{\vec{r}}(\vec{u}_{\vec{r}}) \tag{9.12}$$

where

$$p_{\vec{r}}(\pm 1) = \frac{1}{2}(1 \pm m_{\vec{r}}) \quad \text{(Ising)},$$

and

$$P(\{S_{\vec{r}}\ ;\ \vec{u}_{\vec{r}}\}) = \prod_{\vec{r}\in\Lambda} P_{\vec{r}}(S_{\vec{r}})q_{\vec{r}}(\vec{u}_{\vec{r}})$$

where

$$p_{\vec{r}}(\pm 1) = \frac{1}{2}(n_{\vec{r}} \pm m_{\vec{r}})\ ,\ p_{\vec{r}}(0) = 1 - n_{\vec{r}} \quad \text{(BEG)}\ .$$

The local equilibrium assumption hypothesizes that the velocity distribution is Gaussian:

$$q_{\vec{r}}(\vec{u}_{\vec{r}}) \propto \exp\left[-\frac{\mu c^2}{2}\frac{(\vec{u}_{\vec{r}} - \vec{\nu}_{\vec{r}})^2}{k_B T}\right] \quad \text{where} \quad \vec{\nu}_{\vec{r}} = \langle \vec{u}_{\vec{r}} \rangle\ . \tag{9.13}$$

Finally, the isothermal assumption is that the temperature $T$ in (9.13) is constant.

We shall now determine the time evolution of the flow by imposing local conservation laws. We wish to evolve

$$n_{\vec{r}}\ ,\ m_{\vec{r}} \quad \text{and the momentum} \quad n_{\vec{r}}\ \vec{\nu}_{\vec{r}}\ ;$$

for simplicity, we shall denote by $\rho_{\vec{r}}$ any one of these quantities. Each time step consists of convection followed by diffusion.

*Convection step*: Change $\rho_{\vec{r}}$ to $\rho^C_{\vec{r}}$ using the local conservation law:

$$\rho^C_{\vec{r}} - \rho_{\vec{r}} + \tau \sum_{i=1}^{6} j^{\rho C}_{\vec{r},\vec{r}+\vec{e}_i} = F_{\vec{r}} \tag{9.14}$$

where $\tau$ is a small time step,

$$j^{\rho C}_{\vec{r},\vec{r}+\vec{e}_i} = \rho_{\vec{r}}(\vec{\nu}_{\vec{r}} \cdot \vec{e}_i) - \rho_{\vec{r}+\vec{e}_i}(\vec{\nu}_{\vec{r}+\vec{e}_i} \cdot \vec{e}_i)\ , \tag{9.15}$$

and $F_{\vec{r}}$ represents the stress forces generated at previous diffusive step; actually, $F_{\vec{r}}$ appears only when $\rho_{\vec{r}}$ is the momentum $n_{\vec{r}}\ \vec{\nu}_{\vec{r}}$. The rule (9.14) is valid only in the bulk; at the boundaries we use the no-slip momentum transfer:

$$\vec{\nu}^{\,C}_{\vec{r}} = (\pm\nu_w, 0, 0) \quad \text{if } \vec{r} \text{ is occupied.}$$

*Diffusive step*: Change the convected $\rho^C_{\vec{r}}$ using the local conservation law:

$$\rho^D_{\vec{r}} - \rho^C_{\vec{r}} + \tau \sum_{i=1}^{6} j^{\rho D}_{\vec{r},\vec{r}+\vec{e}_i} = 0\,, \tag{9.16}$$

where

$$j^{\rho D}_{\vec{r},\vec{r}+\vec{e}_i} = j^{(\rho,0)}_{\vec{r},\vec{r}+\vec{e}_i} - j^{(\rho,0)}_{\vec{r},\vec{r}+\vec{e}_i}\,, \tag{9.17}$$

and

$$j^{(\rho,0)}_{\vec{r},\vec{r}+\vec{e}_i} = \Gamma^\rho \rho^C_{\vec{r}}(1 - n^C_{\vec{r}+\vec{e}_i})\varphi(\langle\Delta H\rangle/k_B T)\,, \tag{9.18}$$

where $\Gamma^\rho$ are microscopic friction coefficients, for simplicity assumed to be identical for $\rho = m, n$ and $n\,\vec{\nu}$, and $\varphi(\lambda)$ is the transition rate; $j^{(\rho,0)}_{\vec{r}+\vec{e}_i,\vec{r}}$ is defined in a similar way (replacing $\vec{r}, \vec{r}+\vec{e}_i$ by $\vec{r}+\vec{e}_i, \vec{r}$ respectively).

Notice that for diffusion to take place by hopping from $\vec{r}$ to $\vec{r}+\vec{e}_i$, the cage at must be vacant, which is accounted by the factor

$$(1 - n^C_{\vec{r}+\vec{e}_i})\,.$$

The transition rate $\varphi(\Delta H/(k_B T))$ is determined by the difference in the Hamiltonian at two neighboring cages. The so called local "detailed balance" rule dictates that $\varphi$ satisfies: $\varphi(\lambda) = e^{-\lambda}\varphi(-\lambda)$, thus ensuring dissipative relaxation towards equilibrium.

To close the iterative time-step we define

$$n'_{\vec{r}} = n^D_{\vec{r}}\,,\quad m'_{\vec{r}} = m^D_{\vec{r}}\,, \tag{9.19}$$

$$\vec{\nu}\,'_{\vec{r}} \cdot \vec{e}_i = \vec{\nu}^{\,D}_{\vec{r}} \cdot \vec{e}_i + \frac{1}{n^C_{\vec{r}}}\left(j^{nD}_{\vec{r},\vec{r}+\vec{e}_i} - j^{nD}_{\vec{r},\vec{r}-\vec{e}_i}\right) \tag{9.20}$$

and the components of $F_{\vec{r}}$ by

$$F^i_{\vec{r}} = \sum_{k=1}^{6} j^{n\nu^i D}_{\vec{r},\vec{r}+\vec{e}_k}\,. \tag{9.21}$$

Note that the second term on the right-hand side of (9.20) is the velocity due to diffusion in accordance with Fick's law.

We now take the new quantities $n_{\vec{r}}$, $m_{\vec{r}}$, the momentum and force as defined in (9.19)–(9.21), and proceed with the next time step.

Figure 9.7 shows the simulation of the above procedure for liquid-liquid wetting: it shows both the total density and the velocity. The computation was done for a system with $L_x = 60$, $L_z = 21$, with periodic boundary condition along $\widehat{x}$, and assuming translational invariance along $\widehat{y}$. BEG model was used with $J = 0.3$, $G = -0.15$, $\Delta_{\vec{r}} = 0.15$ and $h_{\vec{r}} = 0$ in the Hamiltonian in (9.11). The latter choice assures a static contact angle of 90°. The wall dimensionless velocity is 0.0035. Only one half of the simulated system around one of the two liquid/liquid interfaces is shown at the steady state. Figure 9.8 shows the horizontal velocity for the first four layers. The numbers inside the figures are values of the order parameter.

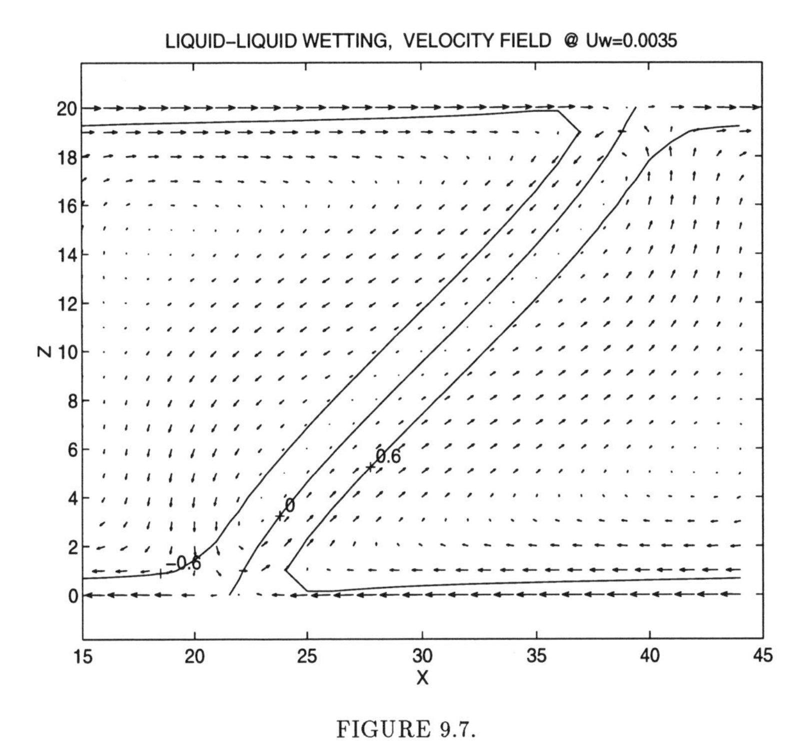


FIGURE 9.7.

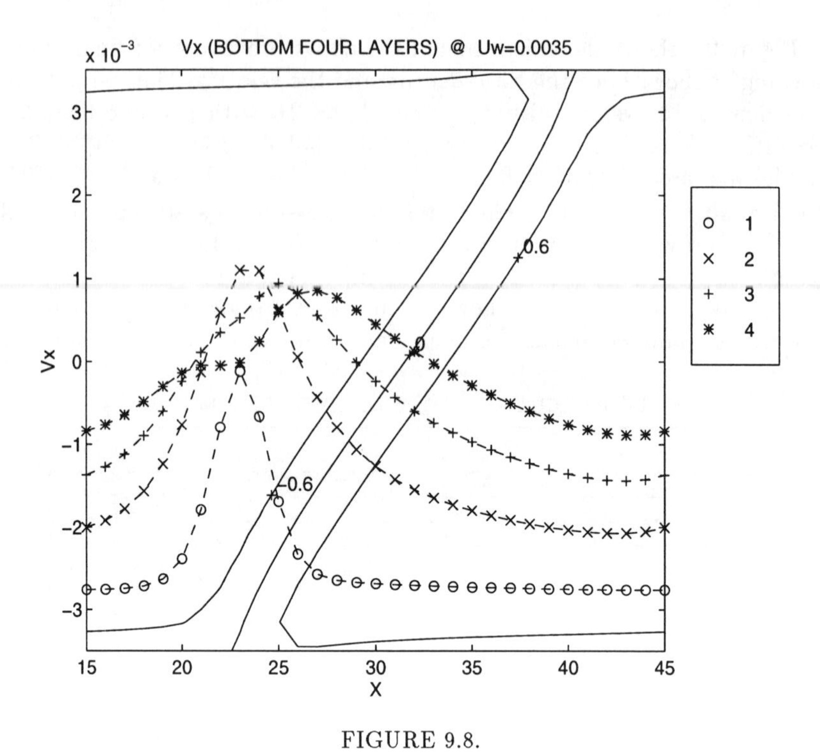


FIGURE 9.8.

Notice the total velocity slip in the first layer adjacent to the wall, the opposite circulating flows on both sides of the interface and the tank-tread velocities tangential to the interfaces, similar to the $MD$ results of Ref [13].

## 9.4 Future directions

The above model has been extended by A. Khan and Y. Shnidman to spinodal decomposition in sheared channel. High-temperature one-phase fluid undergoes temperature quench. This results in a process of nucleation on growth leading towards phase separation. Using the convective-diffusive BEG model Khan and Shnidman simulated the time evolution of this process under shear, including hydrodynamic and wetting effects.

The convective-diffusive lattice-gas model incorporates convective and diffusive steps with feedback, for each time-step. The physical conditions imposed are quite transparent. However the model is limited by the mean-field and isothermal local equilibrium assumptions. It needs extensive testing by comparing to solutions obtained by other methods and to experiments.

Possible future generalizations: (i) Relax the isothermal constraint to allow heat conduction and temperature evolution; (ii) include surfactant interaction.

Another important question can be stated as follows:

*Problem (1).* Can one pass to a continuous limit in the convective-diffusive lattice-gas model (by making the lattice spacing go to zero)?

If such a limit model is possible, it would be interesting to compare it with Model $H$ (derived, for instance, in [15, Chap. 8]). This model is based on continuum conservation laws for order parameter and momentum, and it includes noise terms:

$$\frac{\partial \phi}{\partial t} = \lambda \nabla^2 \frac{\delta H}{\delta \phi} - w \nabla \phi \cdot \frac{\delta H}{\delta \vec{g}} + \zeta_\phi \ ,$$

$$\frac{\partial g_i}{\partial t} = (\delta_{ij} - \nabla_i \nabla_j / \nabla^2) \left[ \eta \nabla^2 \frac{\delta H}{\zeta_{g_j}} - w(\nabla_j \phi) \frac{\partial H}{\delta \phi} + \zeta_{g_j} \right] ;$$

here

$$H = \int \left[ \frac{1}{2} \ r\phi^2 + \frac{1}{2} \ |\nabla \varphi|^2 + u\phi^4 + \frac{1}{2} \ g^2 - h\phi - \vec{h}_g \cdot \vec{g} \right] dxdydz,$$

$1/\nabla^2$ is the inverse Laplacian, $\zeta_\phi$, $\zeta_{g_j}$ are noise terms, $\vec{g}$ is the momentum, $\phi$ is the order parameter, $\delta H / \delta \phi$ is the integrand in the Euler–Lagrange first variation expression, $\delta H / \delta g_j$ is similarly defined, and $\lambda, \eta, w$ are phenomenological parameters.

## 9.5 References


[1] A. Friedman, *Mathematics in Industrial Problems*, IMA Volume 16, Springer–Verlag, New York (1988).

[2] Y. Shnidman, *Forced wetting in a driven diffusive lattice-gas model*, preprint.

[3] E.B. Dussan V., *The moving contact line: The slip boundary condition*, J. Fluid Mech., 77 (1976), 665–684.

[4] S.F. Kistler and L.E. Scriven, in *"Computational Analysis of Polymer Processing"*, (J.R.A. Pearson and S.M. Richardson, Eds.), Chap. 8, pp. 248–299, Applied Science Publishers, London (1983).

[5] C.C. Ngan and E.B. Dussan V., *On the dynamics of liquid spreading on solid surfaces*, J. Fluid Mech., 209 (1989), 191–226.

[6] T.D. Blake and K.J. Ruschak, *Wetting: static and dynamic contact lines, in Liquid Film Coating: Scientific Principles and their Technological Implications*, P.M. Schweizer and S.F. Kistler, eds., Chap 3, Chatman and Hall, London (1996).

[7] A. Friedman and J.J.L. Velázquez, *The analysis of coating flows near the contact line*, J. Diff. Eqs., 119 (1995), 137–208.

[8] Y.D. Shikhmurzaev, *The moving contact line of a smooth solid surface*, Int. J. Multiphase Flow, 19 (1993), 589-910.

[9] Y.D. Shikhmurzaev, *A two-layer model an an interface between immiscible fluids*, Physica A. 192 (1993), 47–62.

[10] T.D. Blake and J.M. Haynes, *Kinetics of liquid/liquid displacement*, J. Coll. Inter. Sci., 30 (1969), 421–423.

[11] H. Eyring, *Viscosity, plasticity, and diffusion as examples of absolute reaction rates*, J. Chem. Phys., 4 (1936), 283–291.

[12] J.I. Frenkel, *Kinetic Theory of Liquids*, Oxford University Press, Oxford (1946).

[13] P.A. Thompson and M.O. Robbins, *Simulations of contact-line motion: Slip and dynamic contact angle*, Phys. Rev. Lett., 63 (1989), 766–769.

[14] A. Friedman, *Mathematics in Industrial Problems, Part 6*, IMA Volume 57, Springer–Verlag, New York (1993).

[15] P.M. Chaikin and T.C. Lubensky, *Principles of Condensed Matter Physics*, Cambridge University Press, New York (1995).

# 10

# Multi-scale problems in modeling semiconductor processing equipment

Modeling semiconductor processing equipment is progressing rapidly in many directions including advanced methods for surface-to-surface radiative exchange and plasma discharges, and refined capabilities to describe thermal reactive flows. Simultaneous with these equipment modeling advances there has been steady improvement in modeling the physics and chemistry occurring on the microscopic scale typical of integrated circuit features. A primary goal when applying these models to actual process equipment is to elucidate the relationship between the controllable, macroscopic characteristics of the process (e.g., pressure, temperature, chemical composition) and the microscopic feature characteristics one is trying to control (e.g., film conformality, critical feature size, local temperature field). The coupling of these so called "equipment-scale" and "feature-scale" models poses a serious computational problem since the length-scale varies as much as seven orders of magnitude between them. On February 16, 1996 Erik W. Egan from Motorola (Advanced Products Research and Development Lab, Austin) described a few real-world examples in which process control on the wafer scale (10 cm), die-scale (5–10 mm), pattern-scale (10–100 $\mu m$) and feature-scale ($0.1 - 1\mu m$) may all be necessary and interrelated. He proposed a simplified, prototypical multi-scale model which might be amenable to mathematical analysis.

## 10.1 The multi-scale problem

Silicon wafers today have diameters of 100–200 mm; 300 mm is under development. A wafer consists of die parts of diameter 5–10 mm. Within a die there are pattern variations of length-scale 10's–100's $\mu m$. Each such pattern has finer features of size $0.2 - 1.0\mu$. This scale progression is illustrated in Figure 10.1.

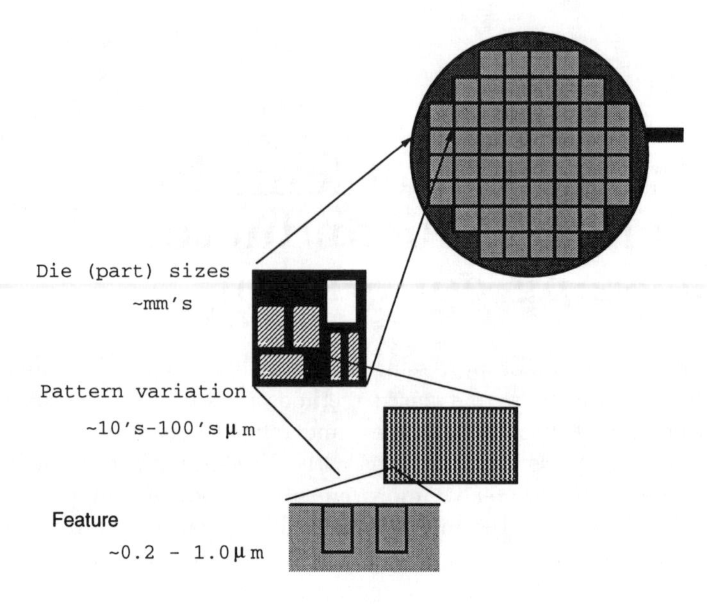


FIGURE 10.1.

A single wafer is processed in a tube of cross-section diameter up to several times larger than the diameter of the wafer, and of length 10's of cm. A multi-wafer tool has similar cross-section diameter, but its length is hundreds of cm; Figure 10.2 shows schematically such a tool.

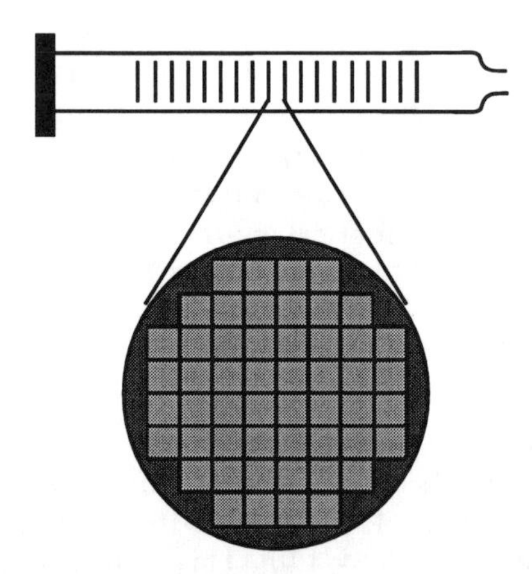

FIGURE 10.2.

Figure 10.3 describes schematically a specific tool used for rapid thermal processing (RTP). The chamber is divided into two parts by a quartz window. A series of lamps are arranged in one part, and the wafer is set up in the other part. The wafer is heated very rapidly by the lamps – the temperature is increased as much as 100°C/sec for about 10 seconds, until it reaches in excess of 1000°C. The temperature is held there for up to a few 10's of seconds before the wafer is allowed to cool. During this short duration annealing, oxidation and deposition of thin films may take place.

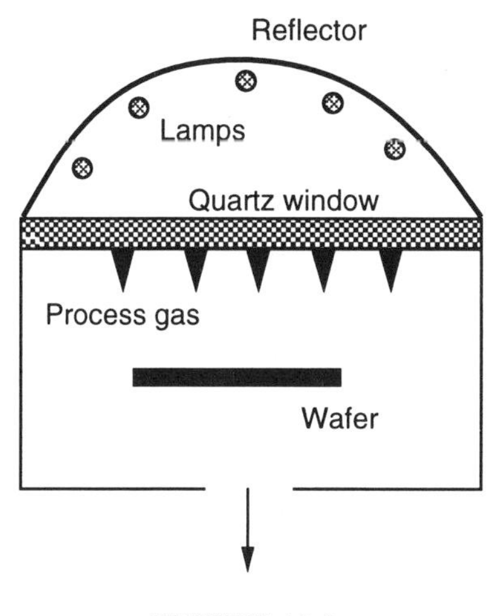


FIGURE 10.3.

The patterned areas of different materials on the wafer have different radiative properties (emissivity, absorptivity). Therefore one is forced to deal with pattern-dependent heat-flux to and from the surface [1]; see Figure 10.4.

*Problem*: How to model transient and steady-state wafer temperature on all relevant scales?

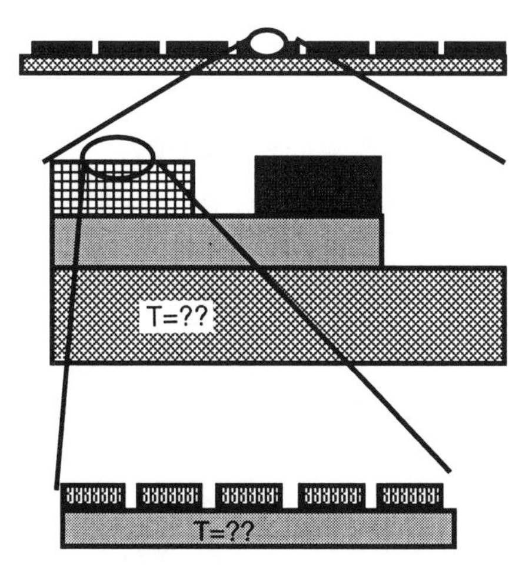


FIGURE 10.4.

The second example is processing by plasma, as schematically shown in Figure 10.5.

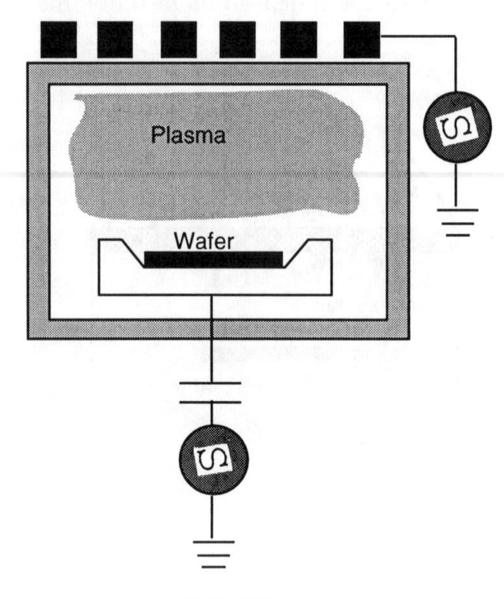


FIGURE 10.5.

Applied electric field excites the plasma. The plasma chemistry then produces reactant species consisting of ions, electrons and excited neutrals. These are transported to the surface of the wafer (by drift and diffusion) where they begin to etch away (by chemical and physical processes) the exposed areas which are not covered by photoresist material; see Figure 10.6. By-products re-deposit or leave the system. It is highly desired to do the plasma etching at low and uniform temperature, so as not to disturb the dopants already implanted in the wafer.

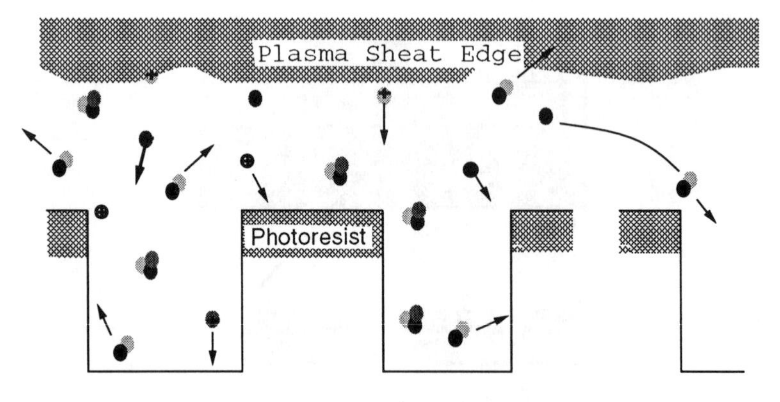


FIGURE 10.6.

The above examples of RTP and plasma etch illustrate problems where the equipment-scale to feature-scale is seven orders of magnitude. The computational approach to-date has been to decouple the problem:

(i) Model the equipment assuming a flat surface of wafer and cross-wafer uniformity in temperature and in etch/deposition rate.

(ii) Use the global solution obtained in (i) as a boundary condition to study the local features, such as the uniformity of deposition and temperature.

Models exhibiting the coupling of the equipment-scale and feature-scale were recently considered by Gobbert, Cale and Ringhofer [2][3] and by Gobbert and Ringhofer [4]. In [5] Gobbert and Ringhofer used the method of matched asymptotic expansion to determine the local solution as a boundary layer in a homogenization problem; this approach was also used in the examples studied in [2][3][4].

## 10.2 Mathematical models

We first consider the RTP model. For simplicity we take the wafer to be a 2-d domain:

$$\Omega = \left\{(x_1, x_2); -a < x_1 < a\ ,\ -b < x_2 < \varepsilon f\left(x_1, \frac{x_1}{\varepsilon}\right)\right\}$$

where

$$f(x_1, \xi_1) \text{ is periodic in } \xi \text{ of period } 1\ ; \tag{10.1}$$

it represents the small features of the wafer, assuming "local" periodicity and just one small-scale feature.

Denote by $E$ the emissivity of the surface and by $A$ its absorptivity ($0 < A < 1$); both account for surface fluxes across the surface of the wafer which is exposed to the lamps, and both vary along the small scale. We denote by $Q_0$ the lamp source, assumed to be constant. Then the temperature $T$ satisfies:

$$\frac{\partial T}{\partial t} = \nabla(\alpha \nabla T) \quad \text{in} \quad \Omega\ , \tag{10.2}$$

$$\frac{\partial T}{\partial n} = E\left(x_1, \frac{x_1}{\varepsilon}\right)\sigma(T^4 - T_\infty^4) - A\left(x_1, \frac{x_1}{\varepsilon}\right)Q_0 \quad \text{on} \quad x_2 = \varepsilon f\left(x_1, \frac{x_1}{\varepsilon}\right)\ , \tag{10.3}$$

$$\frac{\partial T}{\partial n} = 0 \quad \text{elsewhere on} \quad \partial\Omega\ , \tag{10.4}$$

where $n$ is the outward normal, with initial condition

$$T(x_1, x_2, 0) = T_0 \quad \text{(constant)}.$$

Here $\alpha$ is the thermal diffusivity and $\sigma, T_\infty$ are constants.

*Problem (1).* Analyze the solution $T$ for $\varepsilon$ small; in particular, find how

$$\int_\Omega |\nabla T|^2$$

depends on the functions $E$ and $A$.

There is a need in industry to control the temperature in the wafer so as not to upset the doping profile already in place; the temperature variation is usually required to be less than 3°C.

This motivates the problem of minimizing $\int_\Omega |\nabla T|^2$ when the functions $E$ and $A$ are considered to be control variables subject to certain constraints.

The next model we consider is plasma etch. This is a difusive/directional transport with pattern-dependent surface reaction $R_s$; $R_s$ accounts for reactant consumption and etch product generation.

Consider first the case of just one reactant species, and denote its concentration by $c$. The reactant moves in the region

$$\Omega_0 = \{(x_1, x_2); -a < x_1 < a \ , \ \varepsilon f\left(x_1, \frac{x_1}{\varepsilon}\right) < x_2 < b\}$$

where $f(x_1, \xi_1)$ is as in (10.1). The concentration $c$ satisfies:

$$\frac{\partial c}{\partial t} = \nabla(D\nabla c) + q\mu\mathbf{E}\cdot\nabla c + R(c) \quad \text{in} \quad \Omega_0 \ , \tag{10.5}$$

$$c = c_0 \quad \text{on} \quad x_2 = b \ , \tag{10.6}$$

$$\frac{\partial c}{\partial n} = 0 \quad \text{on} \quad x_1 = \pm a \ , \tag{10.7}$$

$$\frac{\partial c}{\partial n} = R_s\left(x_1, \frac{x_1}{\varepsilon}, c\right) \quad \text{on} \quad x_2 = \varepsilon f\left(x_1, \frac{x_1}{\varepsilon}\right) \ , \tag{10.8}$$

with initial condition

$$c(x_1, x_2, 0) = c_0 \ . \tag{10.9}$$

Here $D$ is the diffusion coefficient, $\mathbf{E}$ is a known electric field, $R(c)$ is the reaction rate in the bulk, $q$ is the species charge and $\mu$ is its mobility; $f$ is as in (10.1) and $R_s(x_1, \xi_1, c)$ is 1-periodic in $\xi_1$.

*Problem (2).* Analyze the solution $c$ for small $\varepsilon$; in particular, find the concentration near the boundary $x_2 = \varepsilon f(x_1, x_1/\varepsilon)$.

The system (10.5)–(10.9) deals with one species $c$. In the actual plasma etch problem there are several species $c_1, c_2, \ldots, c_N$. They are coupled by reaction rates

$$R_i = R_i(c_1, \ldots, c_N) \quad \text{in the bulk, and}$$

$$R_{is} = R_{is}\left(x_1, \frac{x_1}{\varepsilon}, c_1, \ldots, c_N\right) \text{ on the surface.}$$

*Problem (3)* Extend Problem (2) to the case of $N$ concentrations.

So far we have formulated 2-scale models. As mentioned in §10.1, there are actually 4 different scales in the wafer. (cf. Figure 10.1). This means that instead of $x_2 = \varepsilon f(x_1, x_1/\varepsilon)$ one should take

$$x_2 = \varepsilon f_1\left(x_1, \frac{x_1}{\varepsilon}\right) + \varepsilon^2 f_2\left(x_1, \frac{x_1}{\varepsilon}, \frac{x_1}{\varepsilon^2}\right) + \varepsilon^3 f_3\left(x_1, \frac{x_1}{\varepsilon}, \frac{x_1}{\varepsilon^2}, \frac{x_1}{\varepsilon^3}\right) \tag{10.10}$$

where

$$f_1(x_1, \xi_1),\ f_2(x_1, \xi_1, \eta_1),\ f_3(x_1, \xi_1, \eta_1, \zeta_1) \text{ are 1-periodic in all the variables } \xi_1, \eta_1, \zeta_1.$$

*Problem (4)*. Extend the results of Problems (1)–(3) to the case

$$x_2 = \varepsilon f_1\left(x_1, \frac{x_1}{\varepsilon}\right) + \varepsilon^2 f_2\left(x_1, \frac{x_1}{\varepsilon}, \frac{x_1}{\varepsilon^2}\right) \tag{10.11}$$

$\left(E \text{ and } A \text{ in (10.3) and } R_s \text{ in (10.8) are then functions of } x_1, \frac{x_1}{\varepsilon}, \frac{x_1}{\varepsilon^2}\right)$, and more generally to the case (10.10).

## 10.3 Partial solution

Partial solutions to the problems posed in §10.2 are already available in the literature. To describe these results we shall concentrate, for clarity, on the equation

$$\Delta u^\varepsilon = Q(x) \quad \text{in a domain} \quad \Omega_\varepsilon\ , \tag{10.12}$$

where a variable point in $\Omega_\varepsilon$ is denoted by $x = (x_1, x_2)$. The domain $\Omega_\varepsilon$ is shown in Figure 10.7; its boundary consists of

$$\Gamma_\varepsilon : x_2 = \varepsilon f\left(x_1, \frac{x_1}{\varepsilon}\right)$$

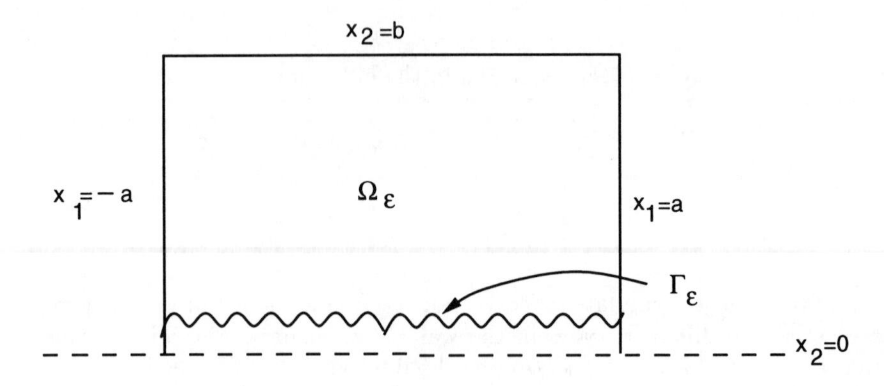


FIGURE 10.7.

and of three line segments: $x_2 = b$ , $x_1 = \pm a$. The function $u^\varepsilon$ satisfies:

$$\frac{\partial u^\varepsilon}{\partial n} + p\left(x_1, \frac{x_1}{\varepsilon}\right) u^\varepsilon = g\left(x_1, \frac{x_1}{\varepsilon}\right) \quad \text{on} \quad \Gamma_\varepsilon \ ,$$
$$u = 0 \quad \text{elsewhere on} \quad \partial\Omega_\varepsilon \ ,$$

where

$$p(x_1, \xi_1) \ , \ g(x_1, \xi_1) \text{ are 1-periodic in } \xi_1 .$$

We define averages

$$P(x_1) = \int_0^1 \sqrt{1 + f_{\xi_1}(x_1, \xi_1)^2}\ p(x_1, \xi_1) d\xi_1 \ ,$$

$$G(x_1) = \int_0^1 \sqrt{1 + f_{\xi_1}(x_1, \xi_1)^2}\ g(x_1, \xi_1) d\xi_1 \ .$$

Denote by $\Omega_0$ the rectangle $\{-a < x_1 < a \ , \ 0 < x_2 < b\}$ and let $u^0(x)$ be the solution of

$$\Delta u^0 = Q \quad \text{in} \quad \Omega_0 \ ,$$
$$\frac{\partial u^0}{\partial n} + Pu^0 = G \quad \text{on} \quad \partial\Omega_0 \cap \{x_2 = 0\} \ ,$$
$$u^0 = 0 \quad \text{elsewhere on} \quad \partial\Omega_0 \ .$$

It is well known (see [6][7]) that

$$u^\varepsilon \to u^0 \quad \text{if} \quad \varepsilon \to 0$$

in the sense that

$$\|u^\varepsilon - u^0\|_{H^1(\Omega_\varepsilon)} \le C\sqrt{\varepsilon} \ . \tag{10.13}$$

Next we want to improve the approximation of $u^\varepsilon$ by adding a boundary layer term in a neighborhood of $\Gamma_\varepsilon$:

$$u^\varepsilon(x) \approx u^0(x) + \varepsilon u^1\left(x, \frac{x}{\varepsilon}\right) . \tag{10.14}$$

Let $\mathcal{D}_{x_1}$ denote the domain in the $\xi = (\xi_1, \xi_2)$ plane bounded by $\xi_1 = 0$ , $\xi_1 = 1$ and lying above the curve

$$\widetilde{\Gamma}_{x_1} : \xi_2 = f(x_1, \xi_1)$$

Let $N_j(x_1, \xi)$ be the solution of

$$\Delta_\xi N_j = 0 \quad \text{in} \quad \mathcal{D}_{x_1} ,$$

$$N_j(x_1, \xi) \to 0 \quad \text{if} \quad \xi_2 \to +\infty ,$$

$$N_j(x_1, \xi) \quad \text{is} \quad \text{1-periodic in} \quad \xi_1 ,$$

and

$$\frac{\partial N_1}{\partial \nu_\xi} = \frac{f_{\xi_1}}{\sqrt{1 + f_{\xi_1}(x_1, \xi_1)^2}} \quad \text{on} \quad \widetilde{\Gamma}_{x_1} ,$$

$$\frac{\partial N_2}{\partial \nu_\xi} = \frac{1}{\sqrt{1 + f_{\xi_1}(x_1, \xi_1)^2}} - \frac{p(x_1, \xi_1)}{P(x_1)} \quad \text{on} \quad \widetilde{\Gamma}_{x_1}$$

where $\nu_\xi$ is the outward normal.

Set

$$u^1(x, \xi) = \left[\frac{\partial u^0}{\partial x_1} N_1(x_1, \xi) + \frac{\partial u^0}{\partial x_2} N_2(x_1, \xi)\right] \tag{10.15}$$

**Theorem 10.1** *There holds:*

$$||u^\varepsilon(x) - u^0(x) - \varepsilon u^1\left(x, \frac{x}{\varepsilon}\right) ||_{H^1(\Omega_\varepsilon)} \leq C\varepsilon . \tag{10.16}$$

A comparison with (10.13) shows that the boundary layer term $\varepsilon u^1(x, x/\varepsilon)$ is an important corrector to the approximation of $u^\varepsilon$ by $u^0$. This corrector represents the coupling between the two scales.

Theorem 10.1 was proved by Belyaev [6] in the special case where $p =$ const., $g = 0$. The proof extends to the general case without difficulty. The method of matched asymptotic expansion used in [5] is another way to arrive at the boundary layer $\varepsilon u^1(x, x/\varepsilon)$, although this approach is just formal.

The method of proof of Theorem 10.1 in [6] easily extends to parabolic equations, such as $\partial u/\partial t = \nabla(\alpha \nabla u) + Q$ and other differential equations.

Consider next the situation with three scales as in (10.11).

A. Friedman, B. Hu and Y. Liu [8] have recently proved that

$$\Bigg\| u^\varepsilon(x) - \Bigg[ u^0(x) + \varepsilon u^1\left(x, \frac{x}{\varepsilon}\right) + \varepsilon \widehat{u}(x) + \varepsilon^2 u^2\left(x, \frac{x}{\varepsilon}, \frac{x_1}{\varepsilon^2}, \frac{x_2 - \varepsilon f_1(x_1, x_1/\varepsilon)}{\varepsilon^2}\right)\Bigg]\Bigg\|_{H^1(\Omega_\varepsilon)} \le C\varepsilon^{3/2} \tag{10.17}$$

where $u^1(x,\xi), u^2(x,\xi,\eta)$ have a structure similar to (10.15); the $u^1$ here depends also on $f_2$. If, in particular, $f_2 \equiv 0$ then it turns out that $u^2 \equiv 0$ and (10.17) yields an improvement over (10.16):

$$\|u^\varepsilon(x) - [u^0(x) - \varepsilon u^1\left(x, \frac{x}{\varepsilon}\right) - \varepsilon\widehat{u}(x)]\, \|_{H^1(\Omega_\varepsilon)} \le C\varepsilon^{3/2}\,. \tag{10.18}$$

Consider a $(C_0\varepsilon)$-neighborhood of the oscillating boundary,

$$\widetilde{\Omega}_\varepsilon = \{x \in \Omega_\varepsilon, \quad \text{distance from } x \text{ to the curve (10.11) is} \le C_0\varepsilon\}\,.$$

The main interest for the semiconductor processing problems described in §10.1 is to determine $u^\varepsilon(x)$ in regions like $\widetilde{\Omega}_\varepsilon$. Since

$$\|\varepsilon\widehat{u}\|_{H^1(\widetilde{\Omega}_\varepsilon)} \le C\varepsilon^{3/2}\,,$$

(10.18) yields

$$\|u^\varepsilon(x) - u^0(x) - \varepsilon u^1\left(x, \frac{x}{\varepsilon}\right)\|_{H^1(\widetilde{\Omega}_\varepsilon)} \le C\varepsilon^{3/2}\,, \tag{10.19}$$

which is a significant improvement over (10.16) in the region of interest $\widetilde{\Omega}_\varepsilon$.

In the same way we anticipate that the $H^1(\widetilde{\Omega}_\varepsilon)$-norm of the expression in the norm on the left-hand side of (10.17) is bounded by $C\varepsilon^2$.

## 10.4 References

[1] J.P. Hell, K.F. Jensen and E.W. Egan, *The potential effect of multilayer patterns on temperature uniformity during rapid thermal processing*, Proc. of the Spring Materials Research Society Meeting, Vol. 387, April 1995.

[2] M.K. Gobbert, T.S. Cale and C. Ringhofer, *One approach to combining equipment scale and feature scale models*, Proceedings of the 187th Meeting of the Electrochemical Society, Reno, Nevada, PV 95–2, May 1995, pp. 553–563.

[3] M.K. Gobbert, T.S. Cale and C. Ringhofer, *The combination of equipment scale and feature scale models for chemical vapor deposition via a homogenization techniques*, 4th International Workshop on Computational Electronics, Tempe, Arizona, November 1995.

[4] M.K. Gobbert and C. Ringhofer, *Misoscopic scale modeling of microloading during LPCVD*, Submitted to the Journal of Electrochemical Society.

[5] M.K. Gobbert and C. Ringhofer, *An asymptotic analysis for a model of chemical vapor deposition on a micro structured surface*, Submitted to SIAM J. Appl. Math.

[6] A.G. Belyaev, *On singular perturbations of boundary problems (Russian)*, Ph. Thesis, Moscow State University, 1990.

[7] O.A. Oleinik, A.S. Shamayev and G.A. Yosifian, *Mathematical Problems in Elasticity and Homogenization*, North-Holland, Amsterdam, 1992.

[8] A. Friedman, B. Hu and Y. Liu, *A boundary value problem for the Poisson equation with multi-scale oscillating boundary*, IMA Technical Report, July 1996.

# 11

# Mathematical modeling needs in optoelectronics

Optoelectronic products are optical products processed by microelectronic techniques. One line of products employs diffractive optics theory to construct various types of diffractive gratings; the underlying mathematical model is that of Maxwell's equations in periodic structure. Another line of products is based on waveguides and it requires, mathematically, the solution of Maxwell's equations in the waveguide with radiation condition at infinity. Both diffractive structures and waveguides have important applications in their own right and will become more significant in the next decade as optoelectronics find its natural role in displays, communications, data storage, and hard copy products. Recently, there have been new developments suggesting that novel structures resulting from the integration of diffractive structures in waveguides can yield resonant reflectors with properties significant for optically active cavities, such as laser diodes. On February 21, 1996 J. Allen Cox from Honeywell Corporation described some of these developments. Additional modeling development will be needed to treat optically active devices, for which the physics requires the inhomogeneous Maxwell equations coupled with the quantum mechanical description of the driven charge carriers.

## 11.1 Diffractive optics and waveguides

Recent progress in diffractive optics has been achieved by mathematical modeling and simulation; we refer to [1, Chap. 14] (and the references therein) where several numerical methods for solving the Maxwell equations in optical periodic structure are described. Some of the outstanding problems are:

(1) For imaging devices, what is an efficient surface relief profile for diffractive element? This is an optimization problem with respect to parameters such as height of profile, shape of profile, wavelength, etc.

(2) What is an efficient method to account for truncated periodic structure, i.e., to what extent the boundary of the device affect the solution computed under the assumption of globally periodic structure?

(3) Incorporate manufacturing and process errors in optimum design codes.

(4) The nonlinear problem: Are grating enhanced nonlinear resonances useful?

Dielectric waveguides are used for gigabit/sec data communication via multimode optical fiber networks. Figure 11.1 shows optical fiber in jacket. The fiber consists of inner part, "core", and outer put, "cladding." The core has index of refraction $n_{co}$ and the cladding has a smaller index of refraction, $n_{c\ell}$. Data is sent through the optical fiber by means of discrete pulses which typically excite multiple eigenvalues of the waveguide. To compute these modes and their power distribution one needs to solve the Maxwell equations (the "direct" problem). The design problem (or the "inverse" problem) is to determine the distribution of excited modes given a measurement of *intensity* distribution in one cross-sectional area of the fiber. (Intensity is proportional to the square of the amplitude.)

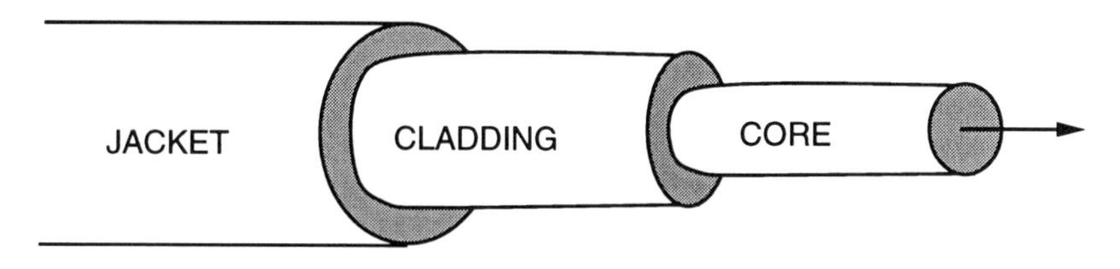


FIGURE 11.1.

Figure 11.2 shows two core profiles; in Figure 11.2(a) the refraction index $n = n_{co}$ is a step function (rectangular waveguide), whereas in Figure 11.2(b) $n = n_{co}$ has a graded (in fact, circular) profile. Single-mode fibers are used in telecommunication whereas multimode fibers are used for data communications (and this is what Honeywell is interested in). As for the dimensions:

For single-mode fiber,

$$2\mu m < \rho < 5\mu m \ ,$$

$$0.8\mu m < \lambda < 1.6\mu m \qquad (\lambda = \text{wavelength}) \ ,$$

$$0.003 < \Delta < 0.01 \ , \quad \text{and} \quad V \sim 1$$

where

$$\Delta = \frac{n_{co}^2 - n_{c\ell}^2}{2n_{co}^2} \ , \ V = \frac{2\pi\rho}{\lambda}(n_{co}^2 - n_{c\ell}^2)^{1/2}$$

($V$ is called the *normalized frequency*);

For multimode fiber,

$$12.5\mu m < \rho < 100\mu m \ ,$$

$$0.8\mu m < \lambda < 1.6\mu m \ ,$$

$$0.01 < \Delta < 0.03 \ , \quad \text{and} \quad V \gg 1 \ .$$

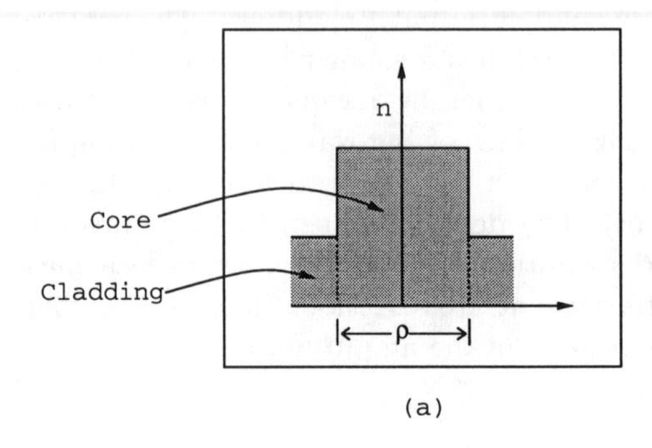


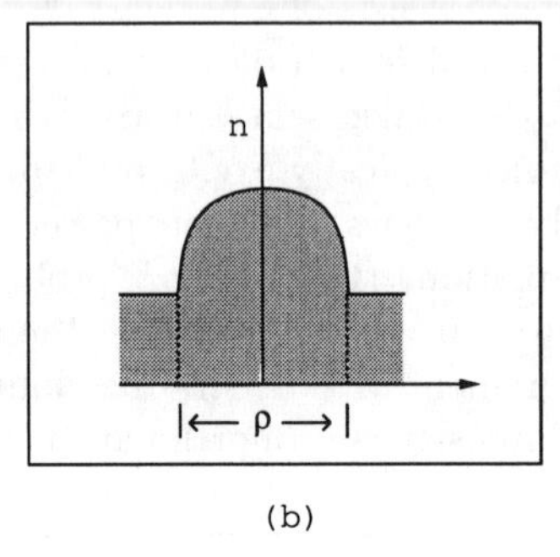


FIGURE 11.2.

For data communications, typical parameters are:

step-index rectangular waveguide, or
graded-index circular fiber as in Figure 11.2(b),

$$\lambda = 0.85\mu m \ ,$$

$$\rho = 25\mu m \ ,$$

$$n_{co} \sim 1.52 \ , \ n_{c\ell} \sim 1.48 \ .$$

## 11.2 The waveguide problem

Figure 11.3(a) shows a waveguide in 3-d; the cladding is contained in a jacket (as shown in Figure 11.1); Figure 11.3(b) shows the profile of the index of refraction.

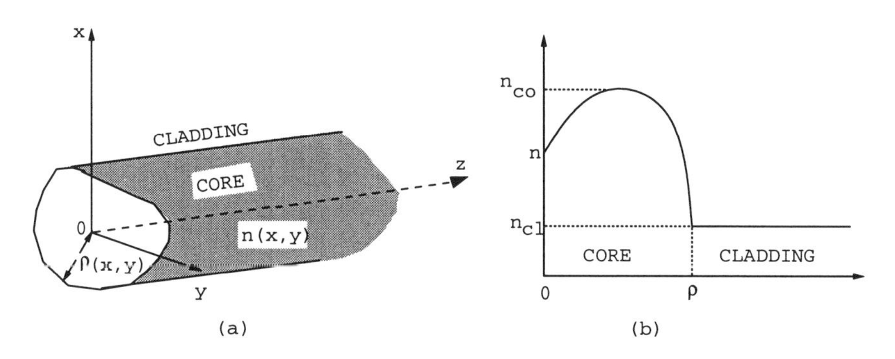


FIGURE 11.3.

The mathematical problem is to solve source-free Maxwell's equations with time harmonic solutions, i.e., solutions

$$e^{-i\omega t}\,\vec{E}\,(x)\ ,\ e^{-i\omega t}\,\vec{H}\,(x):$$

$$\nabla\times\vec{E}-\frac{\omega\mu}{c}\,\vec{H}=0\quad,\quad\mu=1\ ,$$

$$\nabla\times\vec{H}+\frac{\omega\varepsilon}{c}\,\vec{E}=0\quad,\quad\varepsilon=n^2\ .$$

We look for solutions of the form

$$\vec{E}=\vec{e}\,(x,y)e^{i\beta z}\ ,\ \vec{H}=\vec{h}\,(x,y)e^{i\beta z}$$

which are bounded everywhere, and $\to 0$ sufficiently fast as $x^2+y^2\to\infty$; $\vec{e}$ and $\vec{h}$ satisfy the usual continuity conditions at the core/cladding interface; cf. [2, Chap. 3] and [3, Chap. 22]. There is a finite number of bound propagating modes $\{\beta_j\}$, i.e., such that

$$\frac{\omega}{c}\,n_{c\ell}\le\beta_j\le\frac{\omega}{c}\,n_{co}\ ;$$

the $\beta_j$ are eigenvalues, and they depend on the parameter $\omega$. In the core

$$\nabla^2 e+\left[\left(\frac{\omega}{c}\,n_{co}\right)^2-\beta_j^2\right]e=0\ ,$$

and in the cladding

$$\nabla^2 e-\left[\beta_j^2-\left(\frac{\omega}{c}\,n_{c\ell}\right)^2\right]e=0\ .$$

In order to describe some of the waveguide engineering issues, consider a schematic device for data communications, shown in Figure 11.4. The data is generated by laser beam radiating onto rectangular step-index waveguide.

The light beam travels from the rectangular waveguide into graded-index fiber, 100 m – 2000 m, and is received at the other end by rectangular step-index waveguide followed by detector.

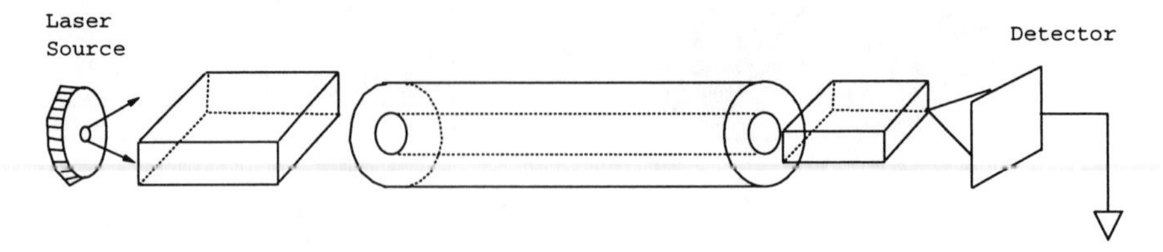


FIGURE 11.4.

The digital data are pulses in uniform, discrete time intervals, shown in Figure 11.5.

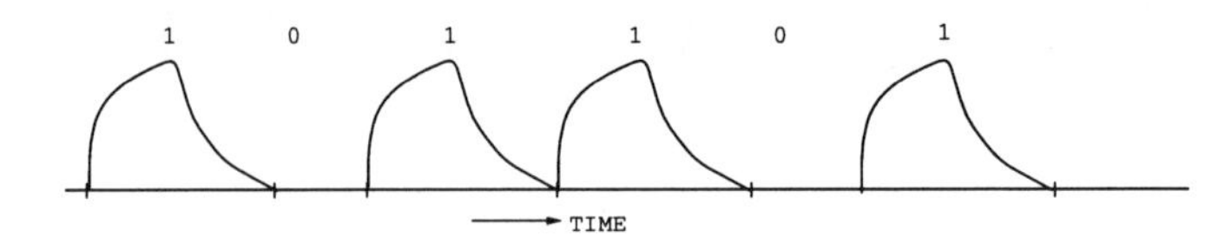


FIGURE 11.5.

Data rate goal is $\sim$ 2 gigabits/sec over 100 m–2000 m. Power in each pulse can be distributed over propagating modes, each moving with *group velocity*

$$v_{gj} = \frac{d\omega}{d\beta_j}$$

(where $\omega, \beta_j$ are defined as above); the group velocity is the velocity by which optical power propagates.

Even if the source is single mode, the propagation is actually multimode. The reasons are:

(i) Launch (input) conditions are not perfect, due to beam intensity irregularity, misalignment, and tilt between the laser source and the rectangular waveguide. Thus other modes are initially excited and they propagate with pulse spreading.

(i) Waveguide perturbations. This is due to nonuniformity in cross-section and refractive index, geometric variations because of bends and twists in the graded-index fiber, and interfaces such as connectors and couplers. All these promote mode coupling.

There is a need for mathematical modeling in waveguides which take into account the multimode which arise as a result of the factors enumerated in (i) and (ii). For the direct problem one needs to determine precisely the launch conditions, mode coupling by perturbations, pulse spreading and transient effects.

The inverse problem is: Given a measure of intensity distribution over a cross-section surface of the waveguide, estimate the power distribution over modes of constant group velocity.

## 11.3 Resonance reflectors for optoelectric devices

Figure 11.6 shows resonant waveguide grating filter. It consists of waveguide grating sandwiched between two dielectric films.

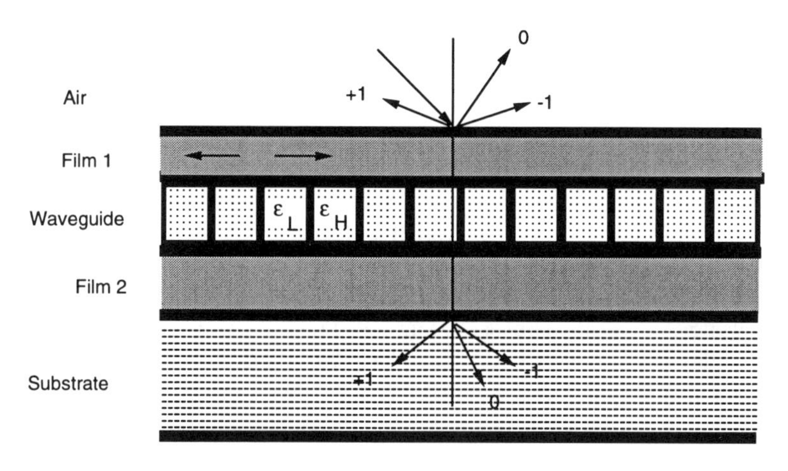


FIGURE 11.6.

Denote by $\Lambda$ the period of the grating and by $\lambda$ the wavelength of plane wave incident to the device. By solving the Maxwell equations one finds that if $\lambda \sim \Lambda$ then the reflectance as a function of the wave length is nearly flat with one sharp peak. Figure 11.7 shows numerical results in case

$$n_L = 1.90 \ , \ n_H = 2.00 \ , \quad n_{\text{substrate}} = 1.52$$

and widths 90 nm of film 1, and 72 nm of the waveguide; here film 2 is absent. For each of the three indicated grating periods, the reflectance peaks up at some particular wavelength, and sharply decays to zero away from these wavelengths. This phenomenon is discussed also in [3, Chap. 7].

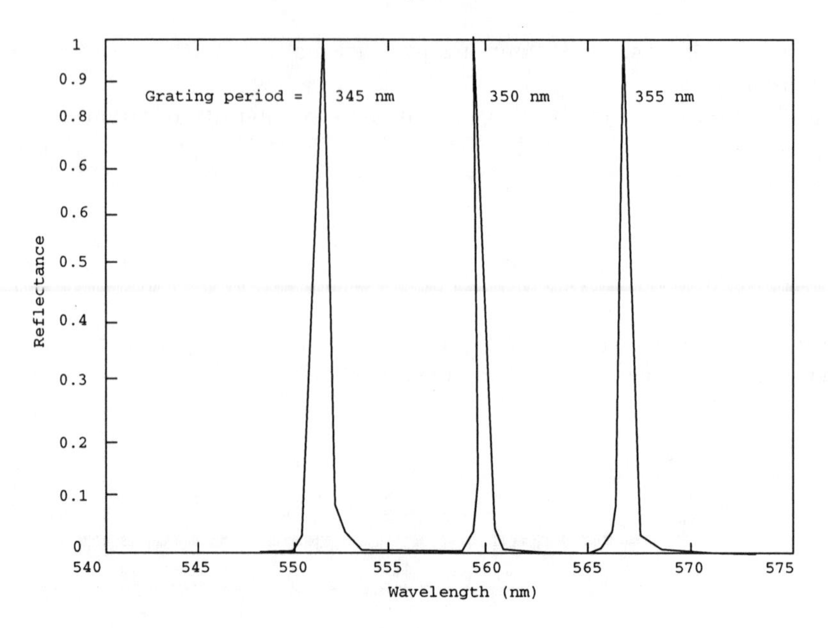


FIGURE 11.7.

Semiconductor laser diode is an example of optically active device; it converts electric power into optical beam (see [2, Chap. 13] and [3]). More complicated devices are described in [3, Chap. 20]. Figure 11.8 shows schematics of optically active device which converts electric power into high intensity electromagnetic radiation in phase, in steady state. The driving source is a steady electric current which feeds the semiconductor laser. The active layer in the semiconductor emits laser which is incident to a resonant cavity. The cavity selects one mode which is reflected back into the active layer and stimulates further emission, causing more radiation to go back from the active layer into the resonant cavity.

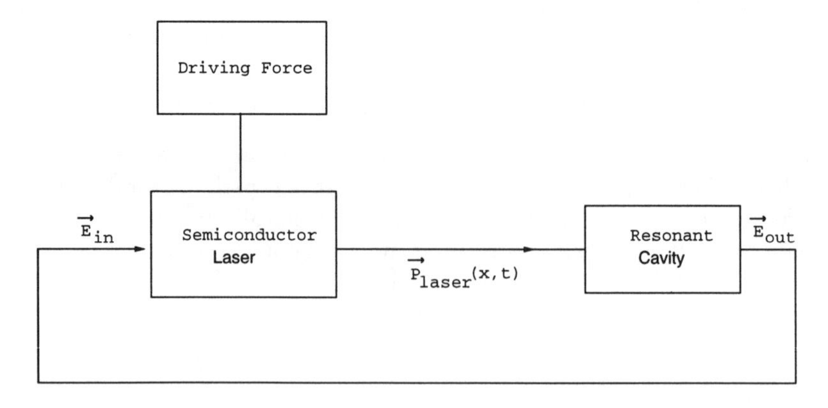


FIGURE 11.8.

The mathematical model consists of the semiconductor laser equations [2, Chap. 13] with sources coming from interaction with the reflected electromagnetic radiation $\vec{E}_{in}$, and of the Maxwell equations in the resonant cavity:

$$\nabla \times \vec{H} = \sigma \vec{E} + \varepsilon \frac{\partial \vec{E}}{\partial t} + \frac{\partial}{\partial t} \vec{P}_{\text{laser}} ,$$

$$\nabla \times \vec{E} = -\mu \frac{\partial \vec{H}}{\partial t} .$$

Honeywell is making such devices, but their modeling and simulation needs to be further developed.

## 11.4 References


[1] A. Friedman, *Mathematics in Industrial Problems, Part 7*, Springer–Verlag, New York (1995).

[2] A. Friedman, *Mathematics in Industrial Problems, Part 3*, Springer–Verlag, New York (1990).

[3] A. Yariv, *Quantum Electronic*, John Wiley & Sons, New York (1988).

# 12

# Modeling solvation properties of molecules

Most practical chemistry and virtually all biological chemistry occur in condensed phases, often in solution, but most molecular modeling techniques treat a molecule in isolation. In spite of this limitation, molecular modeling has yielded much insight and understanding. At the same time treatments of dilute solutions of molecules are becoming more widely available. Because many molecular properties, including reactivities and the structure of biomolecules, depend on solvent, development of reliable techniques are important. Many models of solvation exist, built on increasingly complex descriptions of solvent-solute interaction. The simplest models replace the discrete solvent molecules by a dielectric continuum. The problem then reduces to the solution of Poisson's equation:

$$\nabla \cdot \varepsilon(\mathbf{r}) \nabla \Phi = -4\pi\rho$$

where $\rho$ is the charge density of the molecule and $\varepsilon$ is the dielectric constant. The problem is complicated by the complex shape of the molecule and the irregular distribution of molecular charge.

On March 29, 1996 David K. Misemer from 3M Technical Computations described a method that enables rapid, accurate solution of the problem using boundary element techniques and random sampling; this is joint work with Lawrence Pratt, Greg Tawa and P. Jeff Hay from Los Alamos National Laboratory and Krys Zaklika and John Blair from 3M. He concluded with some questions regarding further improvement of the model and of the computational results.

## 12.1 Physical model of solvation

Figure 12.1(a) shows a molecule in vacuum. Figure 12.1(b) shows the same solute molecule surrounded by some of the solvent molecules present in a solution. One major effect of the solvent can be represented by adding a potential $\Delta\Phi(\mathbf{r}_i)$ at the site of each electron in the solute. Here $\Delta\Phi$ is the electrostatic contribution due to polarization of the solvent. Instead of the Hamiltonian $H_0$ of the solute molecule in vacuum we then have the Hamiltonian

$$H = H_0 + \frac{1}{2}\, e \sum_{i-1}^{N} \Delta\Phi(\mathbf{r}_i) \; ; \tag{12.1}$$

the Schrödinger equation is then

$$H\Psi = E\Psi \tag{12.2}$$

where $\Psi$ (= the wave function) depends on all the electrons and $E$ is the energy of the solute in solvent.

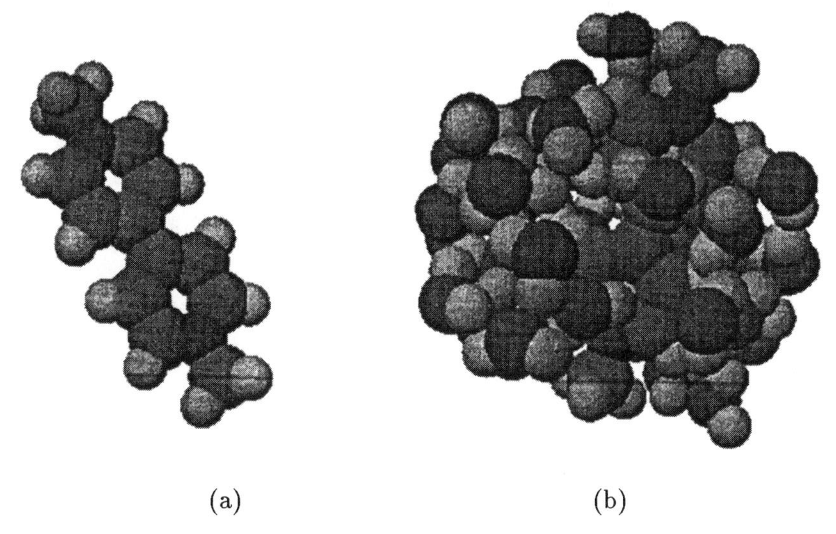

(a) (b)

FIGURE 12.1.

Many properties of molecules in solution are significantly changed, including:

(i) structure of biomolecules and their interaction with other molecules;

(ii) chemical reactivity, and

(iii) color shifts in dye molecules (solvato-chromic effects).

The interaction of the solvent and the solute is complicated and includes hydrogen bonding in addition to electrostatic effects. One must also account for the fact that solvent molecules are excluded from the volume occupied by the solvent molecule. The remainder of the discussion will focus exclusively on electrostatic effects.

The charge distribution within a neutral molecule can give rise to dipole or higher-order moments, which create an electrostatic field. In a polar solvent, composed of molecules with dipole moments, the molecules respond to solute's field and a net polarization of the surrounding medium results. This polarization creates the potential that acts back on the solute molecule. Treating the case where solvent molecules are replaced by simple objects like dipoles, illustrated in Figure 12.2, is still quite demanding.

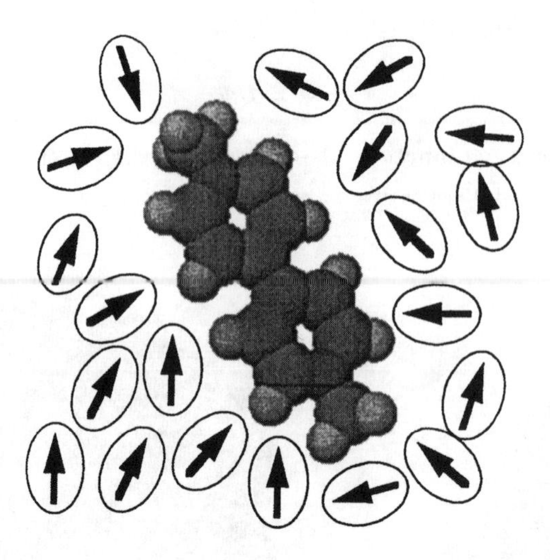

FIGURE 12.2.

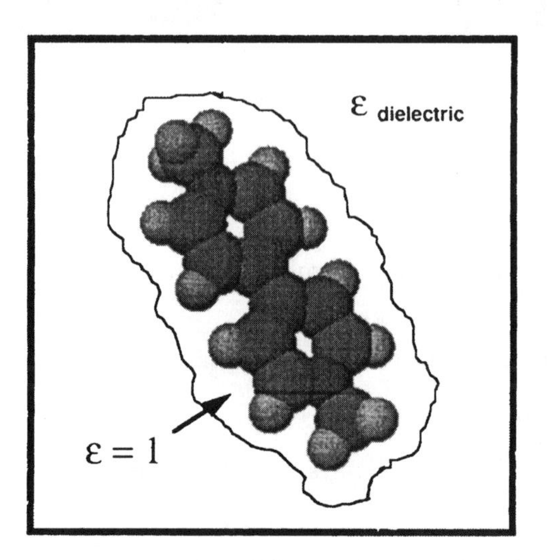


FIGURE 12.3.

An alternate approach is to replace the discrete solvent molecules by a macroscopic dielectric constant, $\varepsilon_{\text{dielectric}}$ characteristic of the solvent. This approach is known as the *continuum solvent approach*. We refer to the recent review by Cramer and Truhlar [1] which describes how this approach is complemented with various models of the solute molecule itself and its interaction with the dielectric bulk of the solvent. For water $\varepsilon_{H_2O} = 80$, so pronounced effects are possible. Figure 12.3 also shows the cavity created by the solute in the dielectric bulk; the dielectric constant in the cavity is

$\varepsilon = 1$. The dielectric continuum model of a solvent provides a well-defined problem amenable to analysis.

## 12.2 Mathematical description

The potential $\Phi$ satisfies

$$\nabla \cdot (\varepsilon \nabla \Phi(x)) = -4\pi\rho(x) \ , \ x \in \mathbb{R}^3 \tag{12.3}$$

where $\rho$ is the charge density due to the solute (as opposed to the polarization charge in the solvent); $\rho = 0$ outside the solute. Denote by $\Omega$ the cavity created by the solute in the bulk, and set

$$\varepsilon = \begin{cases} 1 & \text{in} \quad \Omega \\ \varepsilon^+ & \text{in} \quad \mathbb{R}^3 \backslash \overline{\Omega} \ , \end{cases}$$

$$\Phi = \begin{cases} \Phi^- & \text{in} \quad \Omega \\ \Phi^+ & \text{in} \quad \mathbb{R}^3 \backslash \overline{\Omega} \ . \end{cases}$$

Then (12.3) can be rewritten as

$$\begin{aligned} \Delta\Phi^- &= -4\pi\rho \quad \text{in} \quad \Omega \ , \\ \Delta\Phi^+ &= 0 \quad \text{in} \quad \mathbb{R}^3 \backslash \overline{\Omega} \ , \end{aligned} \tag{12.4}$$

plus the boundary conditions

$$\Phi^- = \Phi^+ \quad \text{on} \quad S \ , \tag{12.5}$$

$$\frac{\partial \Phi^-}{\partial n} = \varepsilon^+ \frac{\partial \Phi^+}{\partial n} \quad \text{on} \quad S \tag{12.6}$$

where $S$ is the boundary of $\Omega$. In addition, we impose the condition

$$\Phi(x) \to 0 \quad \text{if} \quad |x| \to \infty \ ;$$

this implies that $|\Phi(x)| \leq C/|x|$ as $|x| \to \infty$.

Set $\Omega^- = \Omega$ , $\Omega^+ = \mathbb{R}^3 \backslash \overline{\Omega}$ and denote by $n^-(x)$ and $n^+(x)$ the normals at $x \in S$ which are pointing outward to $\Omega^-$ and to $\Omega^+$, respectively. The relation (12.3) means that

$$\int_{\mathbb{R}^3} \varepsilon \nabla \Phi(y) \cdot \nabla \psi(y) dy = 4\pi \int_{\Omega} \rho(y)\psi(y) dy$$

for any function $\psi$ in $C^1(\mathbb{R}^3)$ which decays at $\infty$ sufficiently fast so that the volume integral converges. Take, in particular,

$$\psi(y) = \zeta_\delta(y) G(x, y) \ , \quad G(x, y) = \frac{1}{|x - y|}$$

for fixed $x$ in $\Omega^-$, where

$$\zeta_\delta(y) = h\left(\frac{|x-y|}{\delta}\right) \ , \ h(t) = 1 \quad \text{if} \quad t > 1 \ , = 0 \text{ if } t < \frac{1}{2} \text{ and } h'(t) \geq 0 .$$

Then, as $\delta \to 0$ we obtain

$$\int_{\mathbb{R}^3} \varepsilon \nabla\Phi(y) \cdot \nabla_y G(x,y) dy = 4\pi \int_{\Omega} \rho(y) G(x,y) dy \ . \tag{12.7}$$

By integration by parts and the fact that

$$\nabla_y^2 G(x,y) = -4\pi\delta(x-y)$$

we get

$$\int_{\Omega^-} \nabla\Phi^-(y) \cdot \nabla_y G(x,y) dy = \int_{\partial\Omega^-} \Phi^-(y) \frac{\partial G(x,y)}{\partial n^-(y)} dS_y + 4\pi\Phi^-(x) - 4\pi\Phi_0(x)$$

where

$$\Phi_0(x) = \int_{\Omega} G(x,y)\rho(y) dy \ ; \tag{12.8}$$

note that $\Delta\Phi_0 = -4\pi\rho \quad \text{in} \quad \Omega$ .

Similarly

$$\varepsilon^+ \int_{\Omega^+} \nabla\Phi^+(y) \cdot \nabla_y G(x,y) dy = \varepsilon^+ \int_{\partial\Omega^+} \Phi^+(y) \frac{\partial G(x,y)}{\partial n^+(y)} dS_y \ ,$$

since $\Phi^+$ and $G$ decay at $\infty$. Using this in (12.7), and recalling (12.5), we get

$$4\pi\Phi^-(x) + (\varepsilon^+ - 1) \int_{\partial\Omega^+} \Phi^+(y) \frac{\partial G(x,y)}{\partial n^+(y)} dS_y = 4\pi\Phi_0(x) \tag{12.9}$$

Since $\Phi^+(y)$ and $G(x,y)$ are harmonic in $\Omega^+$ and decay to zero at $\infty$,

$$\int_{\partial\Omega^+} \Phi^+(y) \frac{\partial G(x,y)}{\partial n^+(y)} \ dS_y = \int_{\partial\Omega^+} \frac{\partial\Phi^+(y)}{\partial n^+(y)} \ G(x,y) dS_y \ .$$

Using this and (12.6) in (12.9), we obtain the integral representation for $\Phi^-$ in $\Omega^-$:

$$\Phi^-(x) = \frac{\varepsilon^+ - 1}{4\pi\varepsilon^+} \int_{\partial\Omega^-} G(x,y) \frac{\partial\Phi^-(y)}{\partial n^-(y)} dS_y + \Phi_0(x) \ , \ x \in \Omega^- \ . \tag{12.10}$$

Taking $\partial/\partial\ell$ as $x$ converges to a point on $\partial\Omega^-$ and $\ell$ converges to the direction of the outward normal at that point, and using the jump relation for surface potentials, we get

$$\frac{\partial\Phi^-(x)}{\partial n^-(x)} = 2\pi\frac{\varepsilon^+ - 1}{4\pi\varepsilon^+}\frac{\partial\Phi^-(x)}{\partial n^-(x)} + \frac{\varepsilon^+ - 1}{4\pi\varepsilon^+}\int\frac{\partial G(x,y)}{\partial n^-(x)}\frac{\partial\Phi^-(y)}{\partial n^-(y)}dS_y + \frac{\partial\Phi_0(x)}{\partial n^-(x)}\ ,\quad x\in\partial\Omega^-\ . \tag{12.11}$$

Let us introduce the electric fields

$$\mathbf{E} = -\nabla\Phi^-\ ,\ \mathbf{E}_0 = -\nabla\Phi_0 \quad \text{in} \quad \Omega^-\cup S$$

and set

$$E_n(x) = \mathbf{E}\cdot n^-(x)\ ,\ E_{on}(x) = \mathbf{E}_0\cdot n^-(x)\ ,$$

$$n_S(x) = n^-(x) \quad \text{on} \quad S\ .$$

Then we can rewrite the integral equation (12.11) in the form:

$$\frac{\varepsilon^+ + 1}{2\varepsilon^+}E_n(x) = E_{on}(x) + \frac{\varepsilon^+ - 1}{4\pi\varepsilon^+}\int\limits_S[n_S(x)\cdot\nabla_x G(x,y)]E_n(y)dS_y\ . \tag{12.12}$$

We note that the $\Delta\Phi(\mathbf{r}_i)$ in (12.1) is the difference $\Phi^- - \Phi_0$. Thus in order to work with the model (12.1), (12.2) we first need to solve the integral equation (12.12). Before proceeding to describe results based on the integral equation approach, we mention the approach of Kirkwood–Onsager [2][3] (see also [1]). It asserts that

$$\Phi(R) = \sum_{n=1}^{N}\frac{q_n}{|\mathbf{R}-\mathbf{R}_n|} + \sum_{n=0}^{\infty}\sum_{m=-n}^{n}B_{mn}R^nP_n^m(\cos\theta)e^{im\varphi} \tag{12.13}$$

where the first sum is the potential $\Phi_0$ of the solute and the second sum is a series of spherical harmonics which represents the response of the solvent. Here $q_n$ are the charges in position $\mathbf{R}_n$ of the solute molecule and $N$ is the total number of charges. The potential $\Phi_0$ can be expanded into spherical harmonics,

$$\sum_{n=1}^{N}\frac{q_n}{|\mathbf{R}-\mathbf{R}_n|} = \sum_{n=0}^{\infty}\sum_{m=-n}^{n}\frac{E^{n+1}}{R^{mn}}P_n^m(\cos\theta)e^{im\varphi}\ . \tag{12.14}$$

A serious limitation of this method is that it assumes spherical cavity. Another problem is that the convergence of the multi-pole expansions in (12.14) and in the second sum as the right-hand side of (12.13) is very slow, although this may possibly be improved by the fast multi-pole method of Greengard and Rohklin [4].

## 12.3 Numerical results

The method developed for solving the integral equation (12.12) is the boundary element method. This reduces the problem to a linear system of equations

$$[W]E_n = E_{on} \tag{12.15}$$

where $[W]$ is the (non-sparse) matrix

$$W_{ij} = \frac{\varepsilon^+ + 1}{2\varepsilon +} \delta_{ij} - (1 - \delta_{ij}) \frac{\varepsilon^+ - 1}{4\pi\varepsilon^+} n_{\mathbf{R}_s} \cdot \nabla_{\mathbf{R}_s} G(\mathbf{R}_i, \mathbf{R}_j) \Delta S_j$$

and $E_n, E_{on}$ are vectors with components $E_{nj}, E_{onj}$. Here $n_{\mathbf{R}_s}$ is the same as $n_S$ (in (12.12)) and $\mathbf{R}_s$ is a variable point on $S$. The $E_{onj}$ are not computed directly from (12.9). In fact, in order to reduce the computational time they compute $E_{on}$ at just $N$ points $\mathbf{R}_s^i$ and then introduce "effective" or "pseudo"-charges $Q_n$ by solving the $N$ equations

$$E_{on}(\mathbf{R}_s) = -n_{\mathbf{R}_s} \cdot \nabla_{\mathbf{R}_s} \sum_{n=1}^{N} \frac{Q_n}{|\mathbf{R}_n - \mathbf{R}_s|} \quad \text{at} \quad \mathbf{R}_s = \mathbf{R}_s^i \ .$$

Finally they take

$$E_{onj} = -n_{\mathbf{R}_s \cdot \mathbf{R}_s} \sum_{n=1}^{N} \frac{Q_n}{|\mathbf{R}_n - \mathbf{R}_s|} \quad \text{at} \quad \mathbf{R}_s = \mathbf{R}_j \ .$$

The numerical procedure consists of the following steps:

(1) Define molecular surface as union of the van der Waal's spheres at the atomic centers. (The radii of these spheres are documented in the literature.)

(2) Generate atomic pseudo-charges $Q_n$ as described above.

(3) Generate "sub-random" set of points to define the boundary elements.

(4) Solve the linear system for $E_n$, and then compute $\Phi$.

We refer to Chap. 7 in [5] for the definition of sub-random (or Sobol) selection. It produces faster convergence of integrals than does the method of random (or pseudo-random) selection; whereas pseudo-random selection produces convergence rate of $1/\sqrt{n}$, sub-random selection produces convergence rate of $(\log n)/n$.

Figure 12.4 shows comparison between the two methods as applied to the solution of (12.12) by the boundary element method (12.15). The larger black dots are obtained by the sub-random selection, whereas the smaller black dots are obtained by the pseudo-random selection. The sub-random procedure convergences faster.

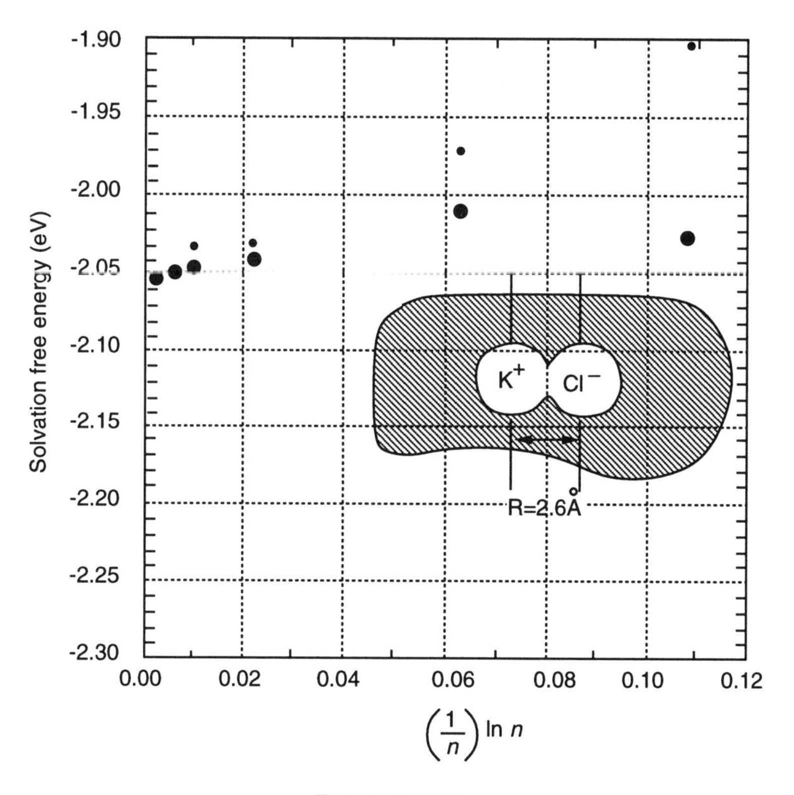


FIGURE 12.4.

We note that the solvation free energy is given

$$W = \frac{1}{2} \sum \Phi(\mathbf{r}_i) q(\mathbf{r}_i)$$

or, for continuous distribution $\rho$ of charges, by

$$W = \frac{1}{2} \int \Phi \rho$$

Since the dielectric displacement $\mathbf{D}$ satisfies $\mathrm{div}\mathbf{D} = \rho$ ,

$$W = \frac{1}{2} \int \Phi \mathrm{div}\mathbf{D} = -\frac{1}{2} \int \nabla\Phi \cdot \mathbf{D} = \frac{1}{2} \int \mathbf{D} \cdot \mathbf{E} \, .$$

The dielectric continuum model is of course not as precise as the discrete solvent model, but its computational time is much smaller. It is therefore desirable to compare numerical results obtained by the two methods. A test case of sodium chloride (table salt) is shown in Figure 12.5, where two graphs describe the solvent energy $W$ as a function of the separation $R$ between the two atoms of the solute.

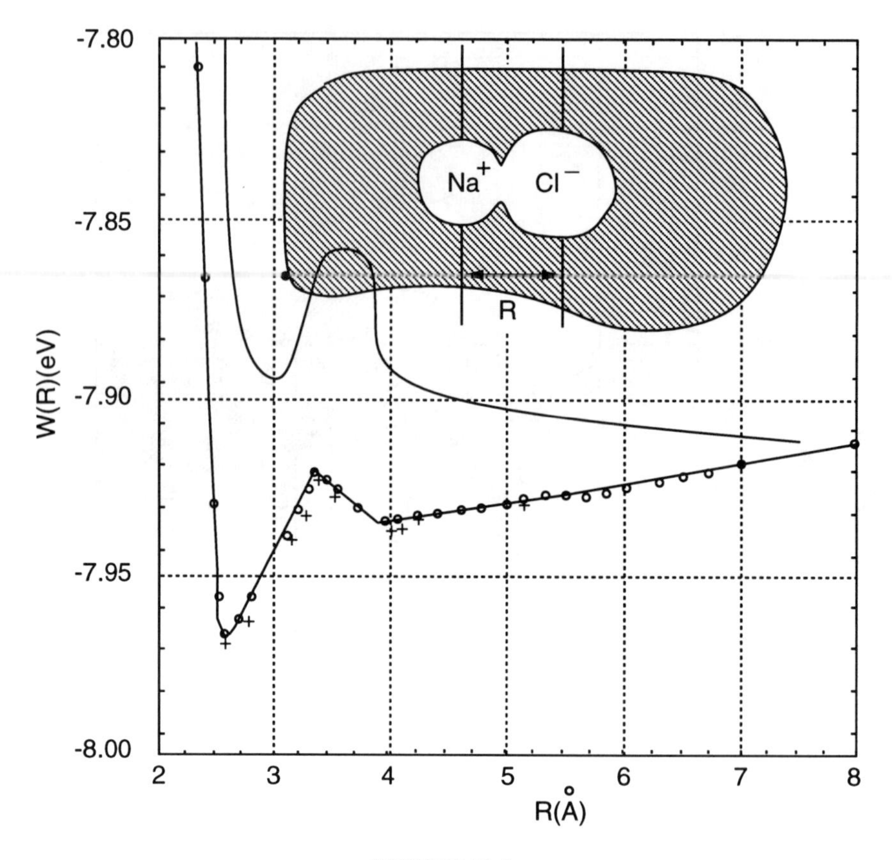


FIGURE 12.5.

The upper curve is obtained by the discrete solvent model; it shows local minimum for $W(R)$ at $R \approx 3\mathring{A}$, but the smallest values of the solvent energy are approached as $R \to \infty$. This is consistent with the fact that sodium chloride dissolves in water. The lower curve in Figure 12.5 is obtained by the dielectric continuum. Here $W(R)$ achieves its absolute minimum at some $R < 3\mathring{A}$ and it increases for $R$ large (behaving like $-e^2/R^2$). Thus the dielectric continuum model is unsatisfactory for large $R$, although it seems quite reasonable for small $R$ (e.g., for $R < 4$). The +'s and $o$'s signify the number of points (boundary elements) on the molecular surface. The $o$'s correspond to 1024; the +'s correspond to 2048. Good convergence is achieved with 1024 points. The solid line connecting the discrete calculations is just a spline fit to the data. The solid line indicates schematically the results from the best molecular calculations involving discrete solvents.

Another test case concerning reactivity has been investigated. Consider for example the situation where $C\ell^-$ comes into contact with a molecule $CH_3C\ell$. At sufficiently high temperature $C\ell^-$ combines with $CH_3C\ell$ and the chlorine atom in $CH_3C\ell$ subsequently becomes free. The process,

$$C\ell^- + CH_3C\ell \Longrightarrow CH_3C\ell + C\ell^- \ ,$$

is shown schematically in Figure 12.6.

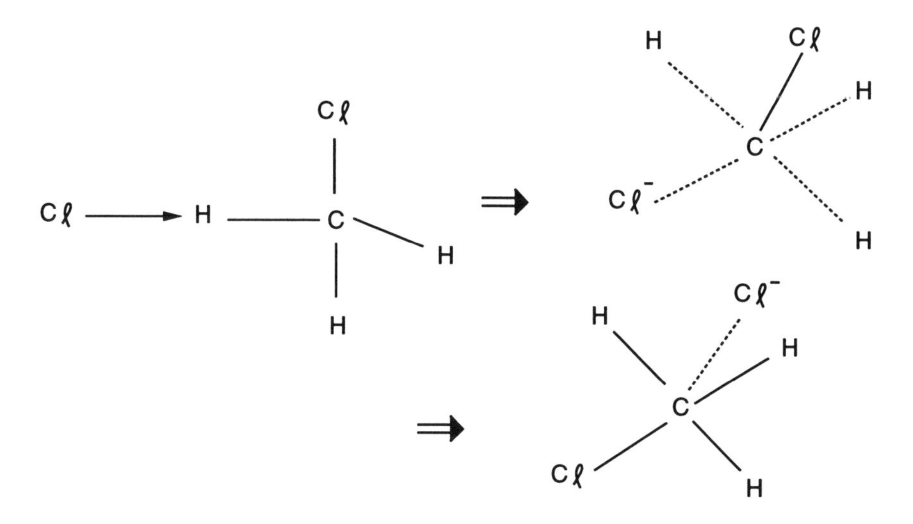


FIGURE 12.6.

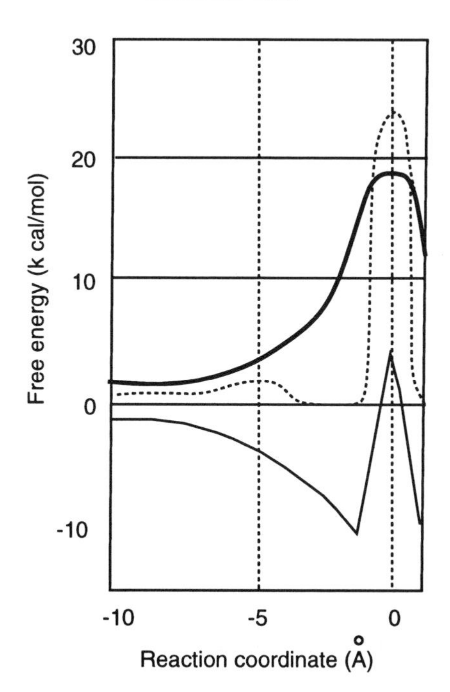


FIGURE 12.7.

The reaction coordinate $R$ is a quantity proportional to the distance from $C\ell^-$ to $C$. We are interested in describing the free energy $W$ as a function of $R$. Figure12.7 shows three curves $W = W(R)$. The bottom one corresponds to no solvent. Note the portion of the curve which is monotone increasing as $R$ goes from −10 to 0 (which is before $C\ell^-$ combines with $CH_3C\ell$). It represents an increase in the free energy (resulting for example, from heating up the molecule); this amount of energy which needed to combine $C\ell^-$ to $CH_3C\ell$ is called the *activation energy*. The next two graphs correspond to the case when the molecule is in solvent: The solid one is computed by the discrete solvent model, and the doted one is computed by the dielectric continuum model. The two reactivity curves have similar features, indicating that the dielectric continuum model is quite reasonable.

## 12.4 Improvements and alternatives

The selection of the random points in the boundary element method is motivated by the fact that the surface of the cavity is not smooth: Intersecting spheres produce non-smooth boundary at the intersection curves. (Boundary element literature in limited to smooth surfaces; it would be helpful to extend the mathematical theory to surfaces with various types of sharpness, or discontinuities of the normals.) Nonetheless a "nearly" uniform selection of the centers of the boundary elements may perhaps do better. Another possible improvement is to use the fast multi-pole method [4], instead of the method of pseudo-charges, for the computation of $E_{onj}$.

Alternative approach altogether is to use a hybrid model; for example, combine the discrete solvable method for a shell around the cavity with the dielectric continuum assumption in the exterior of the shell.

## 12.5 REFERENCES

[1] C.J. Cramer and D.G. Truhlar, *Continuum solvation models: Classical and quantum mechanical implementations*, Chapter 1 in "Reviews in Computational Chemistry," Vol. 6, edited by K.B. Lipkowitz and D.B. Boyd, VCH Publishers, New York (1995), pp. 1–72.

[2] J.G. Kirkwood, *Theory of solutions of molecules containing widely separated charges with special application to Zwitterions*, J. Chem. Phys., 2 (1934), 351–361.

[3] L. Onsager, *Electric moments of molecules in liquids*, J. Amer. Chem. Soc., 58 (1936), 1486–1493.

[4] L. Greengard and V. Rohklin, *A fast algorithm for particle simulations*, J. Comp. Phys., 73 (1987), 325–348.

[5] W.H. Press, B.P. Flannery, S.A. Teukolsky and W.T. Vetterling, *Numerical Recipes*, 2nd edition, Cambridge University Press, Cambridge (1992).

# 13

# Mathematical problems in high-precision interferometric GPS

The Global Positioning system (GPS) is a system of satellites in space used to identify unknown positions on land, sea or in air. The position of the satellites is known at all times. Each satellite sends out signals continuously, and both the signals and their transmission times are precisely controlled by atomic clocks. The current GPS system comprises 25 satellites at heights of 20200 km above the earth. The number of satellites a receiver can see or detect is variable, but occasionally it may be up to 12. In order to identify its position, the receiver produces a reference carrier and beats this carrier against the received signals to extract carrier phase shifts. These phase shifts may be used to obtain millimeter level relative positioning. This scheme is called interferometric GPS. On April 5, 1996 Craig Poling from Lockheed Martin described an application of interferometric GPS to an agricultural navigation system. The objective is to develop on agricultural vehicles (e.g., on combines) a sensor with high precision positioning capabilities in order to manage chemical application correlated to soil map, or real-time soil testing. The mathematical model can be reduced to finding integers $N_1, \ldots, N_k$, called the *ambiguity numbers*. Once these numbers are found, the receiver can determine position with confidence as long as it maintains carrier lock to the satellites. There are many methods to resolve the ambiguity, but each has its limitations. This is an area of current research. A general introduction to the subject and a review of the results obtained up to date can be found in [1]; see also [2]. Some work is being currently done jointly by C. Poling and Brian Leininger (also from Lockheed Martin).

## 13.1 Industrial applications

A group at Lockheed Martin was recently assigned the task to design, develop and field test a prototype precision agricultural navigation system (AGNAV). It was required to achieve 1–3 m row-to-row navigation accuracy, and 10 m absolute accuracy. The agricultural customer felt that communication links from base stations to the agricultural vehicle are unreliable because (i) there are gaps between base stations, and (ii) when the vehicle goes into valleys it loses line of sight with the base station. Traditional differential GPS methods require communication links and hence were not used for AGNAV. The Lockheed Martin project involved the design, building and testing of GPS Interferometric Attitude Sensor. (The *attitude* of a vehicle is defined as the orientation of a specific coordinate system (body fixed frame) with respect to a global reference frame; the three parameters commonly used to define the attitude are the three angles: yaw, pitch and roll). The AGNAV high precision positioning enables managed chemical application correlated with soil map, real-time yield monitoring, and real-time soil testing.

Figure 13.1 shows two antennas 0 and m which are placed on the top of the vehicle in positions $Z_0$ and $Z_m$, and GPS satellite $\ell$ at position $S_\ell$. We denote by $K_{s_\ell}$ the unit vector from $Z_0$ to $S_\ell$; since the distance from the satellite to $Z_0$ is large whereas the distance $d$ from $Z_0$ to $Z_m$ is just a few meters (or even less than 1 m), the unit vector from $Z_m$ to $S_\ell$ is nearly parallel to $K_{s_\ell}$

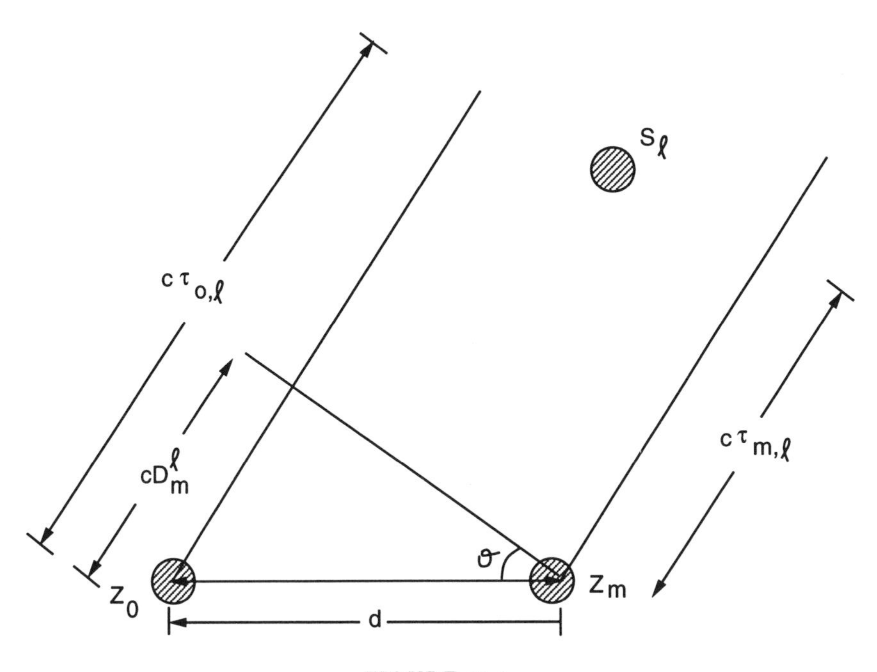


FIGURE 13.1.

Denote by $\tau_{0,\ell}$ the travel time from satellite $\ell$ to antenna 0 and by $\tau_{m,\ell}$ the travel time from satellite $\ell$ to antenna $m$. Then $D_m^\ell = \tau_{0,\ell} - \tau_{m,\ell}$ is the travel time delay between plane wave arrival at antenna $m$ and antenna 0, which will be taken as the reference antenna. If $c$ is the speed of light, then

$$cD_m^\ell = (Z_m - Z_0) \cdot K_{S_\ell}$$

In developing interferometric GPS one has to keep in mind that not only the observer (e.g., the vehicle) is moving but also the satellite is moving; the satellite's velocity is $\approx 3$ miles/sec or 0.01 degree per second.

## 13.2 Interferometric GPS

The GPS signal from satellite $\ell$ at $S_\ell$ to the antenna 0 at location $Z_0$ is

$$S_0^\ell(t) = \mathrm{Re}\big[A\widetilde{x}_{s_\ell}(t + t_{s_\ell} - \tau_{0,\ell})e^{i2\pi(f_c+f_d^\ell)(t+t_{s_\ell}-\tau_{0,\ell})}\big] \ .$$

Here, the signal from $S_\ell$ is actually

$$\widetilde{x}(t + t_{s_\ell})e^{i2\pi(t+t_{s_\ell})}$$

where $t_{s_\ell}$ is the satellite clock offset from GPS system time, $A$ is an amplitude factor that accounts for transmission loss and satellite and receiver antenna patterns, $f_c$ is the carrier or receiver frequency (which depends on the antenna design), and $f_d^\ell$ is the doppler shift; $f_0^\ell \equiv f_c + f_d^\ell$ is then the doppler shifted received frequency at $Z_0$. We shall be working only with the phase part of the received signal:

$$\varphi^{(0,\ell)}(t) = 2\pi(f_c + f_d^\ell)(t + t_{s_\ell} - \tau_{0,\ell}) = 2\pi f_0^\ell(t + t_{s_\ell} - \tau_{0,\ell}) \ . \tag{13.1}$$

The receiver at $Z_0$ produces a reference carrier and it beats this carrier against the received signal. The reference carrier is

$$r_0(t) = \mathrm{Re}[e^{i2\pi f_0(t+t_0)}] \ ,$$

and its phase is

$$\varphi^{(0)}(t) = 2\pi f_0(t + t_0) \ . \tag{13.2}$$

Here $t_0$ is the oscillator clock offset from GPS system time of the receiver at location $Z_0$, and $f_0$ is the reference frequency produced by this receiver.

The beat phase produced by the receiver at location $Z_0$ is then

$$\begin{aligned}\varphi_b^{(0,\ell)}(t) = \varphi^{(0,\ell)}(t) - \varphi^{(0)}(t) &= -2\pi f_0^\ell\tau_{0,\ell} + 2\pi f_0^\ell t_{s_\ell} \\ &\quad -2\pi f_0 t_0 + 2\pi(f_0^\ell - f_0)t \ .\end{aligned}$$

Assuming that $f_c \approx f_0$ and neglecting the doppler shift $f_d^\ell$, we get

$$\varphi_b^{(0,\ell)}(t) = -2\pi f_c\tau_{0,\ell} + 2\pi f_c(t_{s_\ell} - t_0) \ ; \tag{13.3}$$

Note that $\tau_{0,\ell}$ is a function of $t$, since the satellite is moving in time.

When a receiver is switched on at an epoch $T_0$ it measures the instantaneous "fractional" beat phase; the initial integer number $N^{(0,\ell)}$ of cycles between satellite $S_\ell$ and the receiver at $Z_0$ is unknown. When tracking is continued without loss of lock, the integer ambiguity $N^{(0,\ell)}$ remains the same. Writing

$$\varphi_b^{(0,\ell)}(t) = -\Delta\varphi_b^{(0,\ell)}\big|_{T_0}^{t} - 2\pi N^{(0,\ell)} , \tag{13.4}$$

the receiver measures the quantity $\Delta\varphi_b^{(0,\ell)}$, called the *fractional phase* at epoch $t$ augmented by $2\pi$ times the number of integer cycles since the initial epoch $T_0$; the quantity $\Delta\varphi_b^{(0,\ell)}\big|_{T}^{t}/2\pi$ is known as the integral plus fractional part of the Integrated Carrier Phase (ICP).

Setting

$$\varphi_{0,\ell}(t) = \Delta\varphi_b^{(0,\ell)}\big|_{T_0}^{t}$$

we then have from (13.3), (13.4),

$$\varphi_{0,\ell}(t) = 2\pi f_c\tau_{0,\ell}(t) + 2\pi f_c(t_0 - t_{s_\ell}) - 2\pi N^{(0,\ell)} . \tag{13.5}$$

The function $\varphi_{0,\ell}(t)$ is measured, but $N^{(0,\ell)}$ and $\tau_{0,\ell}(t)$ are unknown. The quantities $t_0$ and $t_{s_\ell}$ are also unknown, but we can eliminate them by using two receivers and two satellites, as will be explained below.

First we use, in addition to receiver 0, also receiver $m$. We define the "first differences"

$$\begin{aligned}\varphi_m^\ell(t) &= \varphi_{m,\ell}(t) - \varphi_{0,\ell}(t)\\ &= 2\pi f_c(\tau_{m,\ell}(t) - \tau_{0,\ell}(t)) + 2\pi f_c(t_m - t_0) - 2\pi(N^{(m,\ell)} - N^{(0,\ell)}) ;\end{aligned}$$

substituting

$$D_m^\ell(t) = \tau_{0,\ell}(t) - \tau_{m,\ell}(t)$$

we get

$$\varphi_m^\ell(t) = -2\pi f_c D_m^\ell(t) + 2\pi f_c(t_m - t_0) - 2\pi(N^{(m,\ell)} - N^{(0,\ell)}) . \tag{13.6}$$

Next we define "double differences" by using two satellites, say 0 and $\ell$:

$$\begin{aligned}\varphi_m^0(t) - \varphi_m^\ell(t) &= -2\pi f_c D_m^0 + 2\pi f_c(t_m - t_0) - 2\pi(N^{(m,0)} - N^{(0,0)})\\ &\quad -[-2\pi f_c D_m^\ell + 2\pi f_c(t_m - t_0) - 2\pi(N^{(m,\ell)} - N^{(0,\ell)}) .\\ &= -2\pi f_c(D_m^0 - D_m^\ell) - 2\pi(N^{(m,0)} - N^{(0,0)} - N^{(m,\ell)} + N^{(0,\ell)}) ,\end{aligned}$$

so that the receiver clock offsets $t_m$ and $t_0$ from GPS system time have disappeared. Setting

$$\varphi_m^{0,\ell}(t) = \varphi_m^0 - \varphi_m^\ell(t) ,$$

$$N_m^{0,\ell} = N^{(m,0)} - N^{(0,0)} - N^{(m,\ell)} + N^{(0,\ell)}$$

we can write

$$\varphi_m^{0,\ell}(t) = -2\pi f_c(D_m^0(t) - D_m^\ell(t)) - 2\pi N_m^{0,\ell} . \tag{13.7}$$

We can still remove the unknown integer ambiguity with triple difference, by looking at

$$\varphi_m^{0,\ell}(t^1) - \varphi_m^{0,\ell}(t^2) ,$$

but this introduces measurements at different times, $t^1$ and $t^2$, which has the disadvantage that the position of the receiver, as well as of the satellite position have changed during the time difference; this might defeat the goal of determining the position of the vehicle in real-time.

In formula (13.7) the quantity $\varphi_m^{0,\ell}(t)$ is measurable whereas $D_m^0(t) - D_m^\ell(t)$ is unknown, but can be expressed in terms of the orientation of the body. Denote by $C_B^{ENU}(t)$ the orthogonal transformation that maps vectors from the body coordinate system into the east-north-up coordinate system. Then

$$\begin{aligned} D_m^\ell(t) &= \tau_{0,\ell}(t) - \tau_{m,\ell}(t) = K_{s_\ell} \cdot (Z_m - Z_0)/c \\ &= K_{s_\ell}^{ENU^T}(t) C_B^{ENU}(t)(Z_m - Z_0)^B / c \end{aligned}$$

where superscript "$T$" indicates the transpose and superscript "$B$" indicate the component of a vector in body coordinates. Hence

$$D_m^\ell(t) - D_m^0(t) = (K_{s_\ell}^{ENU^T}(t) - K_{s_0}^{ENU^T}(t)) C_B^{ENU}(t)(Z_m - Z_0)^B / c .$$

Setting

$$ICP^{(m,\ell)}|_{T_0}^t = \Delta\varphi_b^{(m,\ell)}|_{T_0}^t / 2\pi$$

we can then rewrite (13.7) in the form

$$(K_{s_\ell}^{ENU^T}(t) - K_{s_0}^{ENU^T}(t)) C_B^{ENU}(t)(Z_m - Z_0)^B = \lambda_c ICP_m^{0,\ell}|_{T_0}^t + \lambda_c N_m^{0,\ell} \tag{13.8}$$

where

$$ICP_m^{0,\ell}|_{T_0}^t = ICP^{(m,0)}|_{T_0}^t - ICP^{(0,0)}|_{T_0}^t - ICP^{(m,\ell)}|_{T_0}^t + ICP^{(0,\ell)}|_{T_0}^t \tag{13.9}$$

is measured by the receiver; here we used the relation $f_c\lambda_c = c$ where $\lambda_c$ is the wavelength of the carrier. In (13.8) we wish to solve for the unknown body frame $B$ to east-north-up direction cosine matrix $C_B^{ENU}(t)$; the vectors

$$K_{s_\ell}^{ENU}(t) - K_{s_0}^{ENU}(t) , \ (Z_m - Z_0)^B$$

are known: indeed $K_{s_\ell}^{ENU}, K_{s_0}^{ENU}$ are computed from the GPS receiver (Ephemeris) data and the location of the antennas, and $(Z_m - Z_0)^B$ is of course known in the body frame.

In equation (13.8), for each ambiguity $N_m^{0,\ell}$ we can compute $C_B^{ENU}(t)$, but the solution will vary with the choice of $N_m^{0,\ell}$. Only one solution and its corresponding integer ambiguity is the correct solution. How can we find it?

## 13.3 The mathematical challenge

Ambiguity resolution is the process of identifying the correct double difference integer ambiguities $N_m^{0,\ell}$ and corresponding correct solution $C_B^{ENU}(t)$. We recall that this problem makes sense only so long as the receiver maintains carrier lock on the satellite; if it loses carrier lock (due to low signal-to-noise ratio, multipath, etc.) then the integer ambiguities may change, and ambiguity resolution methods are needed to identify a new set of correct integer ambiguities.

The number of feasible integers $N_m^{0,\ell}$ in (13.8) is very large. To illustrate this consider first the case where $m = 2$ and

$$(Z_2 - Z_0)^B = |Z_2 - Z_0| \begin{pmatrix} 0 \\ 1 \\ 0 \end{pmatrix}$$

Then (13.8) becomes

$$|Z_2 - Z_1|(K_{s_\ell}^{ENU^T} - K_{s_0}^{ENU^T}) Col_2 C_B^{ENU} = RHS^{(2,\ell)} \tag{13.10}$$

where $Col_2 C_B^{ENU}$ denotes the second column in $C_B^{ENU}$ and $RHS^{(2,\ell)}$ denote the right-hand side of (13.8).

Since $C_B^{ENU}$ is orthogonal matrix, we easily get

$$|RHS^{(2,\ell)}| \leq 2|Z_2 - Z_0| \ ,$$

or

$$|ICP_2^{0,\ell}|_{T_0}^t + N_2^{0,\ell}| \leq 2\frac{|Z_2 - Z_0|}{\lambda_c} \ . \tag{13.11}$$

The GPS wavelength is $\lambda_c = 19.0293672$ cm (this wavelength is in the range of radar wavelengths). Hence if the receivers were 1.9 meters apart, i.e., 10 wavelengths, then $N_2^{0,\ell}$ could assume the 41 values $-20, \ldots, 0, \ldots, 20$.

To solve for the three components of $Col_2 C_B^{ENU}$ we need to use four satellites: $\ell = 0, 1, 2, 3$. Thus the total number of integer ambiguities is

$$(N_2^{0,1})(N_2^{0,2})(N_2^{0,3}) = 41^3 = 68{,}921 \ .$$

This is the total number of equations, and only one integer value for each of the ambiguities $N_2^{0,1}$ , $N_2^{0,2}$ , $N_2^{0,3}$ is correct, and the value of $Col_2 C_B^{ENU}$ corresponding to the correct integers is the only correct solution.

The ultimate goal is to do interferometric GPS on a global scale with millimeter level positioning. This will require a careful analysis of the GPS signal as it is refracted through the ionosphere and troposphere. The path can be precisely computed by the Fermat principle:

$$\min_{(y(x),z(x))} \int_a^b \frac{\sqrt{1 + y'^2 + z'^2}}{c(x,y,z)} \, dx$$

where $c(x, y, z)$ is the speed of light in the atmosphere at location $(x, y, z)$.

Methods for resolving ambiguities are described in [1, Chap. 9]. There are criteria for ambiguity ranges based on the confidence intervals of the real values of the ambiguities. Correlation of ambiguities is another step in the process of reducing the ambiguities; so is the method of least squares. Reliable, rapid ambiguity resolution methods for large baselines (i.e., large seperation between the antennas) is an area of current research.

A recent collection of articles [2] deals with interferometric GPS methods which rely on the motion of the vehicle.

Interferometric GPS has been used by astronomers to achieve world-wide millimeter level positioning (3–7 mm) and to achieve earth orientation accuracies of $5.6 \times 10^{-8}$ degrees. These accuracies are within a factor of four of those achieved by the most precise VLBI methods using quasars. GPS receiver measurements have recently been used also to perform global ionospheric and tropospheric tomography on the speed of light, that is, determining $c(x, y, z)$ from the travel times.

## 13.4 References

[1] B. Hofmann–Wellenhof, H. Lichtenegger and J. Collins, *GPS Theory and Practice*, 3rd revised edition, Springer–Verlag, Wien New York (1994).

[2] *Understanding GPS Principles and Applications*, E.D. Kaplan ed., Artech House Publishers, Norwood, MA (1996).

# 14

# Questions about the Poisson equation in semiconductor problems

A sufficiently accurate approximation of the steady-state potentials in a reverse-biased silicon semiconductor device can sometimes be obtained with the so-called depletion region approximation. In this approximation, all of the potential drop is assumed to occur in a region that contains only fixed charges (donor and acceptor dopant ions) and that is nearly free of mobile charges (electrons and holes). The potentials are determined by solving the Poisson equation subject to given voltage and zero current boundary conditions on a region whose boundaries have to be calculated as part of the solution. The problem is trivial in 1D, but in 2D and 3D the calculation of the depletion region boundary is much more difficult. On April 19, 1996 Leonard Borucki from Motorola has described the physical setup and the mathematical model. He then proposed several problems whose solution would help in the development of fast algorithm for computing the boundary of the depletion region.

## 14.1 PN junction

A silicon crystal can be made into semiconductor material by introducing impurity atoms. These impurity atoms displace some of the silicon atoms of the crystal. There are two types of impurities: donor and acceptor. Donor impurities are atoms (such as phosphorus or arsenic) that have more valence electrons than are needed to complete the bonds with neighboring atoms of the crystal. When an atom $P$ of phosphorus displaces an atom $Si$ of silicon in the crystal, four of its five valence electrons are used to bond $P$ with its four $Si$ neighbors. This leaves one valence electron, which then moves to the, so called, conduction band and becomes a conduction electron; see Figure 14.1. We call $P$ *donor*, and we denote by $N_d$ the number density of the conduction electrons generated by doping the silicon with $P$. A crystal doped with donors is called $n$-semiconductor.

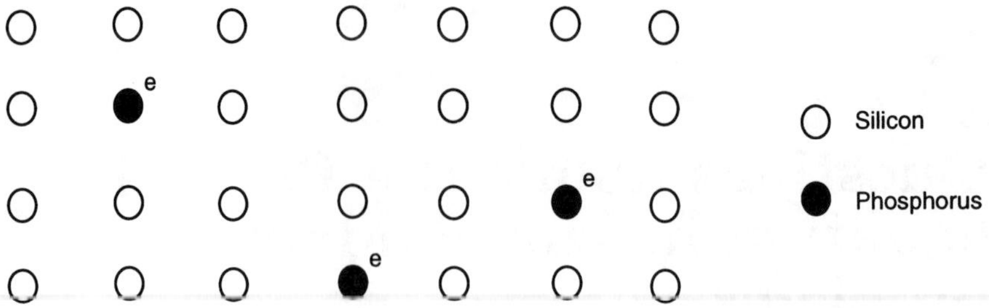

FIGURE 14.1.

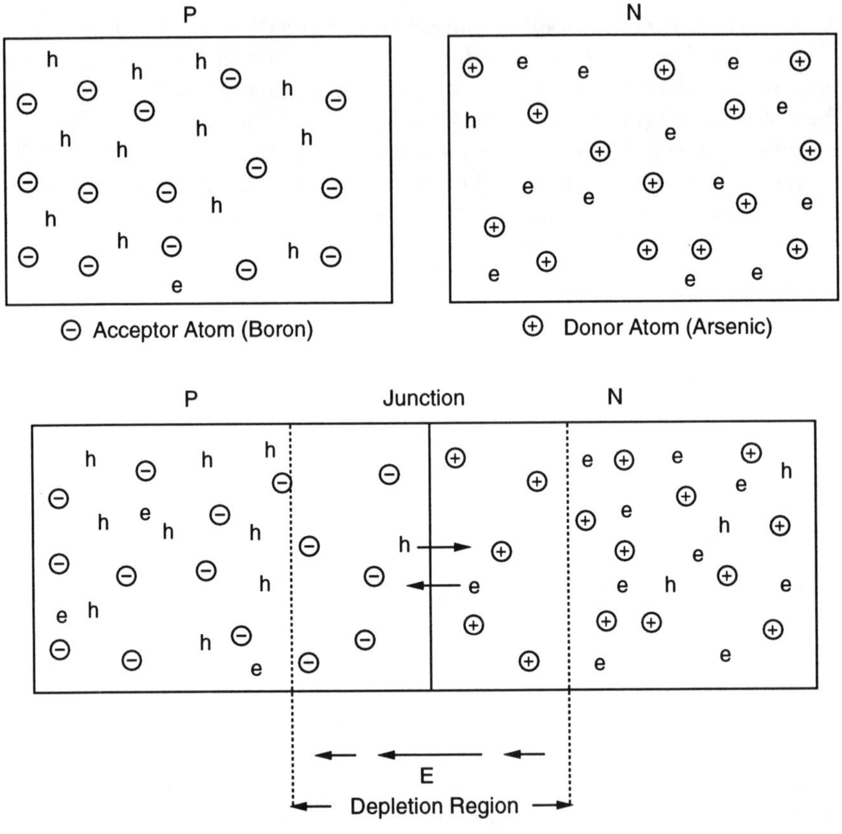

FIGURE 14.2.

Similarly if we implant silicon with boron, one bond is missing, and we refer to it as "hole." This hole can easily be removed by "stealing" nearby valence electron, thus creating a nearby hole. In this way the flow of holes, like the flow of conduction electrons, creates an electric current; electrons have negative charge whereas holes are considered as representing positive charge. We say that boron is *acceptor* and we denote the number density of holes generated by doping with acceptor by $N_a$. A crystal doped with an acceptor is called $p$-semiconductor. For more detail we refer to [1, Chap. 13] and the references therein.

If we bring $p$- and $n$- semiconductors into contact, as shown in Figure 14.2, we get a *pn* junction. The holes from the $p$-domain and the electrons from the $n$-domain that are near the junction diffuse toward each other and create a region called the *depletion* region.

This process gives rise to an electric field $\mathbf{E}$ with potential $\psi$ given by $\mathbf{E} = -\nabla\psi$ which eventually halts the interdiffusion. In the simple $1-d$ case, $\psi = \psi(x)$ and the increase of $\psi$ across the depletion region is called the *built-in potential*; see Figure 14.3

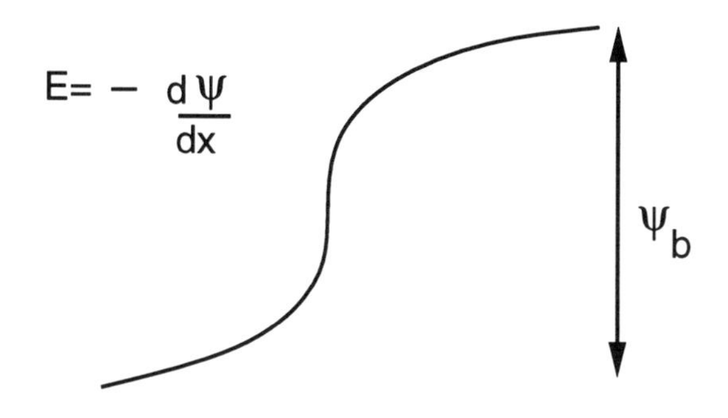


FIGURE 14.3.

The depletion region expands as the applied voltage $V$ increases; see Figure 14.4. When $V$ is increased beyond $\psi_b$ then we speak of a *reverse* biased junction; if $V$ is decreased below $\psi_b$ then the junction is *forward* biased.

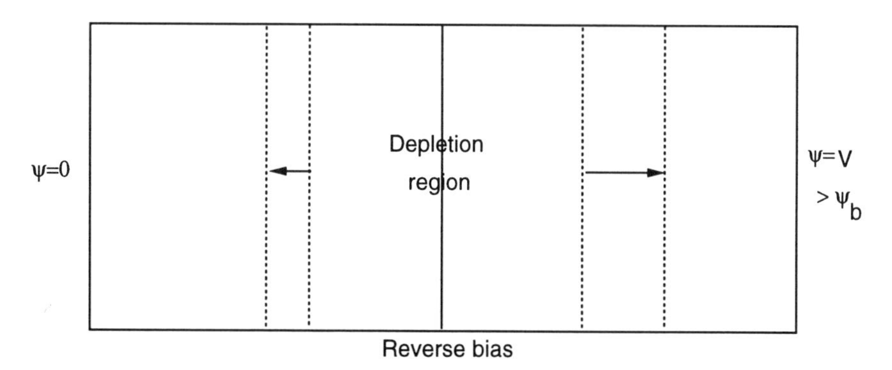


FIGURE 14.4.

Figure 14.5 shows a *pn* diode. It consists of *pn* junction with two lead contacts. If $V$ is decreased (forward bias) then eventually current will begin to flow through the semiconductor. When $V$ is prescribed above a certain level then essentially no current will flow. $NP$ diode is used as "one way street" for current, e.g., as circuit breaker; for more details see [2, Chap. 6] and the references therein.

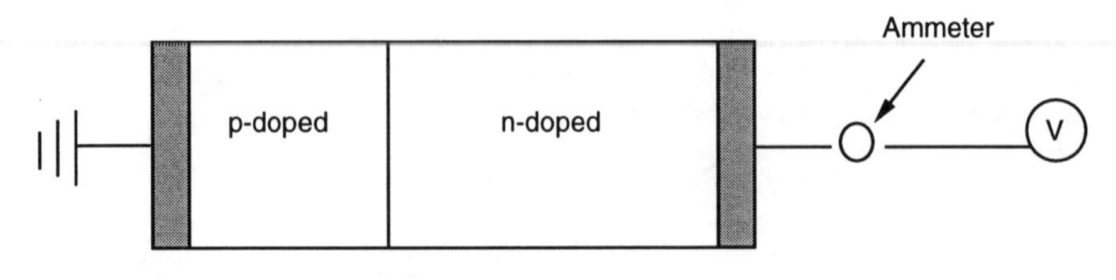


FIGURE 14.5.

Figure 14.6 describes one of the most common semiconductor devices, called a MOSFET transistor. It consists of source and drain *n*-regions in otherwise *p*-domain, a polysilicon gate and oxide spacer. If voltage $V$ is applied at the gate, electrons will move from the source to the drain across the *p*-region and current $I$ will flow. The gray zones in Figure 14.6 are the depletion regions; for more details, see [1, Chap. 13] and the references therein.

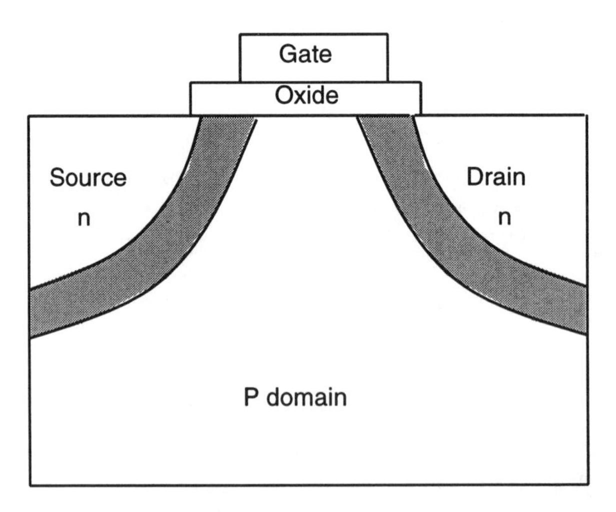


FIGURE 14.6.

In the depletion region there are very few electrons and very few holes. Nevertheless because of the high voltages that are applied to some kinds of devices, (1000 V), the effect of these carriers may be significant, and it occasionally causes avalanche, as illustrated in Figure 14.7: One conduction electron energizes and releases valence electrons from the crystal structure, and as this process continues to grow, it causes breakdown of the junction.

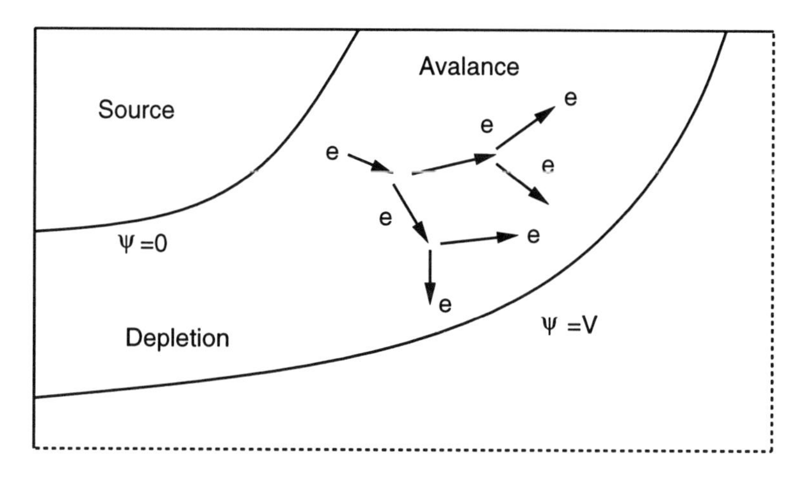


FIGURE 14.7.

The breakdown, caused by the onset of avalanche is illustrated in terms of the I–V curve in Figure 14.7. In order to understand how the onset of avalanche depends on the shape and size of the depletion region, it is important to analyze and compute the depletion boundary as a function of $V$ and of the geometry and material properties of the device.

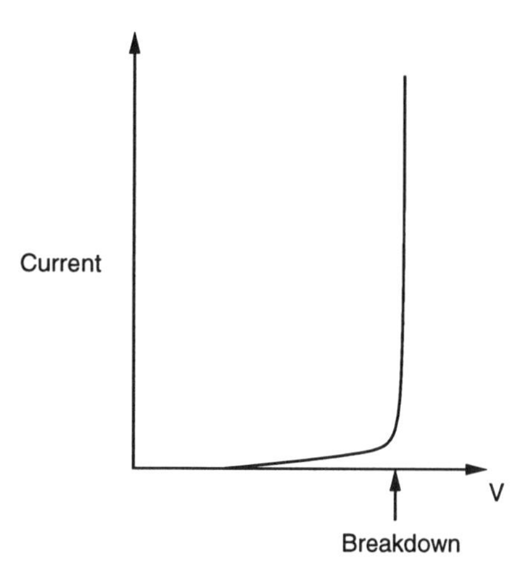


FIGURE 14.8.

## 14.2 Mathematical model

We begin with the device equations:

$$
\begin{aligned}
&\text{div grad } \psi = \frac{q}{\varepsilon}(n - p - C) , \\
&\text{div}\mathbf{J}_n - q\frac{\partial n}{\partial t} = qR , \\
&\text{div}\mathbf{J}_p + q\frac{\partial p}{\partial t} = -qR , \\
&\mathbf{J}_n = qn\mu_n\mathbf{E} + qD_n \text{ grad } n , \\
&\mathbf{J}_p = qp\mu_p\mathbf{E} - qD_p \text{ grad } p
\end{aligned}
\tag{14.1}
$$

where $n$ is the electron concentration, $p$ is the hole concentration, $C = N_d - N_a$ is the net doping concentration, $\mu_n$ and $D_n$ are the mobility and diffusion coefficient for the electrons, and $\mathbf{J}_n$ is the electron flux; $\mu_p$, $D_p$ and $\mathbf{J}_p$ are similarly defined with respect to holes; $q$ is the electron charge, $\varepsilon$ is the permittivity, $E=$ electric field $= -\text{grad}\psi$, and $R$ is the recombination rate; for more details see, for instance, [3].

The depletion approximation model assumes that $n = p = 0$ in the depletion region. In the case of 1-d model the system (14.1) reduces to the following problem [4, Chap. 6]:

$$
\begin{aligned}
&\frac{d^2\psi}{dx^2} = -\frac{q}{\varepsilon}(N_d - N_a) \equiv -\beta \quad \text{if} \quad \alpha_1 < x < \alpha_2 , \\
&\psi = \frac{d\psi}{dx} = 0 \quad \text{at} \quad x = \alpha_1 , \\
&\psi = V , \ \frac{d\psi}{dx} = 0 \quad \text{at} \quad x = \alpha_2
\end{aligned}
\tag{14.2}
$$

where $\alpha_1 \le x \le \alpha_2$ is the depletion region.

Figures 14.9 (a) and (b) show a 2-d geometry with two possible depletion regions.

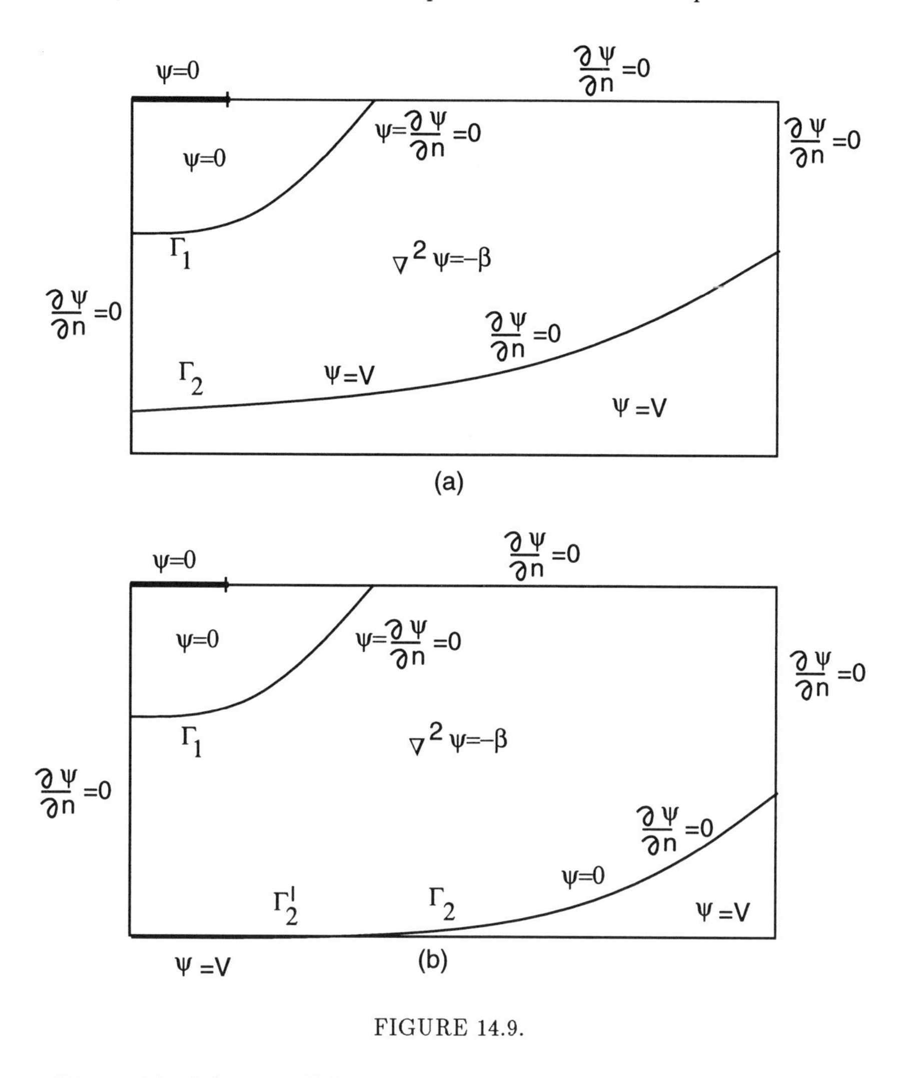


FIGURE 14.9.

Figure 14.10 shows a different 2-d device and a corresponding depletion region.

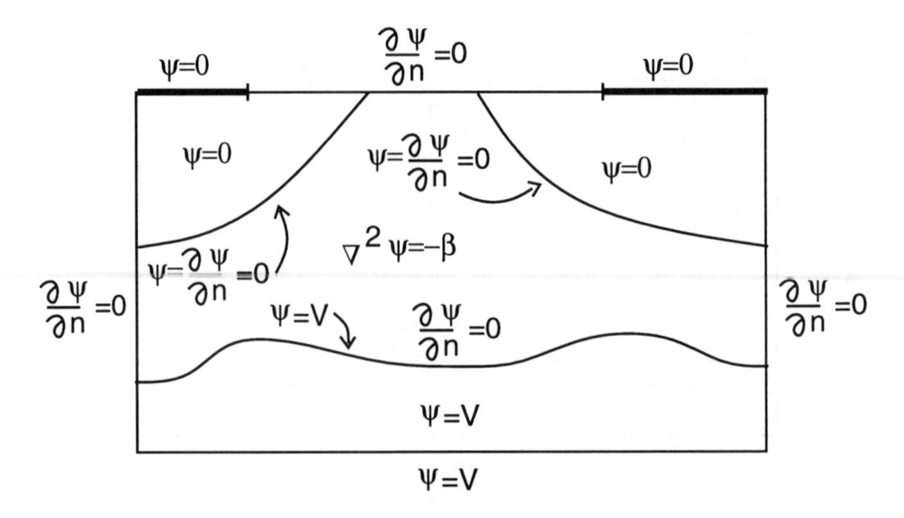


FIGURE 14.10.

In the derivation of the approximate model described in Figures 14.9 and 14.10, the nonlinearities which arise in high field device were ignored. To be more precise one should include their effect, since our interest is primarily in case of high voltage $V$. Nevertheless, the rough model asserted in (14.2) for 1–d or in Figures 14.9, 14.10 (for 2- and 3-dimensions) should be useful for initial understanding.

## 14.3 Questions

The problem of interest to Borucki is to develop an efficient procedure to compute the two boundaries of the depletion region in the geometry of Figure 14.9, or in other geometries, such as in Figure 14.10. He raised the following questions:

*Question (1).* Do the equations describing the depletion region have a unique solution?

*Question (2).* What are the smoothness properties of the boundary of the depletion region?

*Question (3).* Are 0 and $V$ the minimum and maximum of the potential $\psi$?

*Question (4).* If $N_d$ is constant and $N_a$ decreases monotonically with depth and in the lateral directions, are there ever any local extrema in the interior of the depletion region?

These questions are answered in the next section.

## 14.4 Solutions

Consider first the 2-d case and set

$$\Omega_0 = \{(x,y);\ 0 < x < L\ ,\ 0 < y < b\}\ ,$$

$$\ell_0 = \{(x,b); 0 < x < a\}\ ,\ \ell_1 = \{(x,b)\ ;\ a < x < L\}\ ,$$

$$\ell_2 = \{(x,0)\ ;\ 0 < x < L\}\ ,\ Z = (a,b)\ ;$$

$Z$ is the common end-point of $\ell_0$ and $\ell_1$. If we denote the depletion region by $\Omega_0$ then the problem described in Figure 14.9 (a) is the following:

$$-\Delta\psi = \beta \quad \text{in} \quad \Omega_0\ , \tag{14.3}$$

$$\psi = 0\ ,\ \nabla\psi = 0 \quad \text{on} \quad \Gamma_1\ , \tag{14.4}$$

$$\psi = V\ ,\ \nabla\psi = 0 \quad \text{on} \quad \Gamma_2\ , \tag{14.5}$$

$$\frac{\partial\psi}{\partial n} = 0 \quad \text{elsewhere on } \partial\Omega_0 \tag{14.6}$$

where $\Gamma_1$ is the upper boundary of the depletion region and $\Gamma_2$ is the lower boundary. For the problem in Figure 14.9 (b) we denote by $\Gamma_2^1$ the horizontal part of $\partial\Omega_0$ which lies in $\{y = 0\}$ and by $\Gamma_2$ the lower part of the depletion boundary in $\Omega$; then (14.5) is replaced by

$$\begin{aligned} &\psi = V \quad \text{on} \quad \Gamma_2'\ , \\ &\psi = V\ ,\ \nabla\psi = 0 \quad \text{on} \quad \Gamma_2\ . \end{aligned} \tag{14.7}$$

**Theorem 14.1** *If $\beta_x \geq 0$ and $\beta_y \leq 0$ then $0 < \psi < V$.*

Thus the answer to question (3) is yes.

*Proof.* Consider first the geometry of Figure 14.9 (a), i.e., the system (14.3)–(14.6). Differentiating (14.3) in $y$, we get

$$-\Delta\psi_y = \beta_y < 0 \quad \text{in} \quad \Omega_0$$

and then, by the maximum principle,

$$\psi_y \leq 0 \quad \text{in} \quad \Omega_0\ .$$

Similarly $-\Delta\psi_x = \beta_x > 0$ in $\Omega_0$, and

$$\psi_x \geq 0 \quad \text{in} \quad \Omega$$

by the maximum principle. From the inequalities $\psi_y \leq 0$ , $\psi_x \geq 0$ and the boundary conditions $\psi = 0$ on $\Gamma_1$ , $\psi = V$ on $\Gamma_2$ it follows that $0 \leq \psi \leq V$.

In the case of Figure 14.9 (b) we can prove as before that $\psi_x \geq 0$. It follows that $\psi \leq V$ near the bottom of the horizontal boundary $\Gamma_2'$, so that $\psi_y \leq 0$ on $\Gamma_2'$. We can now proceed, as before, to establish that $\psi_y \leq 0$ in $\Omega_0$ (by the maximum principle) so that $0 \leq \psi \leq V$.

From the strong maximum principle we can actually deduce that $\psi_x > 0$ and $\psi_y < 0$ in $\Omega_0$. Thus, the answer to Question (4) is that there are no local extrema in the depletion region; furthermore, $0 < \psi < V$ in $\Omega_0$.

We next assume, in addition to $\beta_x \leq 0$ , $\beta_y \geq 0$, that

$$\beta(x,0) > 0 \ , \ \beta(x,b) < 0 \quad \text{if} \quad 0 \leq x \leq L \ . \tag{14.8}$$

For $V$ large we expect $\Gamma_1$ to be near $\{y = b\}$ so that

$$\Gamma_1 \text{ lies in the set } \{\beta \geq 0\} \ , \tag{14.9}$$

and similarly we expect that

$$\Gamma_2 \text{ lies in the set } \{\beta \leq 0\} \ . \tag{14.10}$$

If we extend $\psi$ by 0 above $\Gamma_1$ and by $V$ below $\Gamma_2$, then

$$-\Delta\psi = 0 \geq \beta \quad \text{above} \quad \Gamma_1 \ ,$$

$$-\Delta\psi = 0 \leq \beta \quad \text{below} \quad \Gamma_2 \ .$$

Thus the extended function $\psi$ satisfies almost everywhere:

$$\begin{aligned} &-\Delta\psi \geq \beta \ , \ \psi \geq 0 \ , \ \psi(\Delta\psi + \beta) = 0 \quad \text{in the set} \quad \Omega \cap \{\psi < V\} \ , \\ &-\Delta\psi \leq \beta \ , \ \psi \leq V \ , (\psi - V)(\Delta\psi + \beta) = 0 \quad \text{in the set} \quad \Omega \cap \{\psi > 0\} \ . \end{aligned} \tag{14.11}$$

These relation show that $\psi$ is a solution of the 2-*sided obstacle problem*, with constant obstacles 0 and $V$. This problem can be given a "weak" formulation as a *variational inequality* [6]:

Let

$$J(\varphi) = \frac{1}{2} \int_\Omega |\nabla\varphi|^2 - \int_\Omega \beta\varphi \ ,$$

$$K = \{\varphi \in H^1(\Omega) \ , \ 0 \leq \varphi \leq V \quad \text{in} \quad \Omega \ , \quad \varphi = 0 \quad \text{on} \quad \ell_0 \ ,$$
$$\varphi = V \quad \text{on} \quad \ell_2\} \ .$$

We seek a solution $\psi$ to the following problem:

$$\psi \in K \ , \qquad J(\psi) = \min_{\varphi \in K} J(\varphi) \ . \tag{14.12}$$

We have shown so far that if $\beta_x \geq 0$, $\beta_y \leq 0$ and (14.9), (14.10) hold, then $\psi$ is a solution of the variational inequality (14.11) and therefore also of (14.12). We now proceed to study problem (14.12) for *general* function $\beta$.

**Theorem 14.2** *For any $\beta$ in $L^\infty(\Omega)$ there exists a unique solution $\psi$ of (14.12), and*

$$\psi \in C^{1+\alpha}(\overline{\Omega}\backslash\{Z\}) ,$$

*for any $0 < \alpha < 1$.*

The uniqueness part answers positively Question (1), assuming that $\beta_x \geq 0$, $\beta_y \leq 0$ and that (14.9), (14.10) hold.

*Proof.* Theorem 14.2 follows from the general theory of the obstacle problem [6, Chap. 1]. The only point that requires an explanation is the regularity at the corner points of $\Omega$. But this follows by reflecting $\psi$ across $x = 0$ or $x = L$, noting that the $\beta$, as extended by reflection, is still a bounded function.

In order to study the regularity of the free boundaries $\Gamma_1, \Gamma_2$ we shall henceforth impose the following conditions:

$$\beta(x, 0) > 0 \ , \ \beta(x, b) < 0 \quad \text{if} \quad 0 \leq x \leq L \tag{14.13}$$

**Theorem 14.3** *If $\beta_y < 0$ in $\overline{\Omega}$ then $\psi_y \leq 0$.*

*Proof.* There are several possible proofs. Consider the penalty approach: We solve

$$\begin{aligned} & -\nabla\psi + \gamma_\varepsilon(\psi) = \beta \quad \text{in} \quad \Omega \ , \\ & \psi = 0 \quad \text{on} \quad \ell_0 \quad \psi = V \quad \text{on} \quad \ell_2 \ , \\ & \frac{\partial\psi}{\partial n} = 0 \quad \text{elsewhere on} \quad \partial\Omega \end{aligned} \tag{14.14}$$

where $\gamma_\varepsilon(t)$ is a penalty function:

$$\begin{aligned} & \gamma_\varepsilon'(t) \geq 0 \ , \\ & \gamma_\varepsilon(t) \to -\infty \quad \text{if} \quad t < 0 \ , \ \varepsilon \to 0 \ , \\ & \gamma_\varepsilon(t) \to +\infty \quad \text{if} \quad t > V \ , \ \varepsilon \to 0 \ , \\ & \gamma_\varepsilon(t) \to 0 \quad \text{if} \quad 0 < t < V \ , \ \varepsilon \to 0 \ . \end{aligned}$$

For convenience we also take

$$\gamma(0) = \min\{\beta\} - 1 \ , \ \ \gamma(V) = \max\{\beta\} + 1 \ .$$

One can show that there exists a unique solution $\psi_\varepsilon$ of (14.14) and, by the maximum principle,

$$0 \leq \psi_\varepsilon \leq V \ .$$

Next,

$$-\Delta \frac{\partial \psi_\varepsilon}{\partial y} + \gamma'_\varepsilon(\psi_\varepsilon) \frac{\partial \psi_\varepsilon}{\partial y} = \frac{\partial \beta}{\partial y} \leq 0 ,$$

and, since $\gamma'_\varepsilon \geq 0$, the maximum principle should yield

$$\frac{\partial \psi_\varepsilon}{\partial y} \leq 0 ; \tag{14.15}$$

since furthermore (by general theory) $\psi_\varepsilon \to \psi$ if $\varepsilon \to 0$, it would then follow that also $\partial\psi/\partial y \leq 0$. There is however a technical difficulty in establishing (14.15) due to the singularity of $\partial\psi_\varepsilon/\partial y$ at the point $Z = (a, b)$.

We shall not give here the details on how to overcome this difficulty, but instead give another proof which is of intrinsic interest. In the functional $J(\psi)$, if we replace the function $\psi(x, y)$ by a monotone decreasing rearrangement $\psi^*(x, y)$, then $\int_\Omega |\nabla\psi|^2$ decreases [6, p. 294] and also $-\int_\Omega \beta\psi$ decreases (since $\beta$ is monotone decreasing in $y$). It follows that $\psi^*$ is also a solution of (14.12) and, by uniqueness, $\psi = \psi^*$, so that $\psi_y \leq 0$.

Introduce the *coincidence set*, in $\overline{\Omega}$,

$$\Lambda = \Lambda_0 \cup \Lambda_V \quad \text{where}$$

$$\Lambda_0 = \{\psi = 0\} , \ \Lambda_V = \{\psi = V\} .$$

From now on we shall also assume, in addition to (14.13), that

$$\frac{\partial \beta}{\partial y} < 0 \quad \text{in} \quad \Omega . \tag{14.16}$$

By the strong maximum principle $\psi_y < 0$ in the non-coincidence set $\Omega_0 = \Omega \backslash \Lambda$. It follows that there exist two curves

$$\widehat{\Gamma}_1 = \{(x, F_1(x)) \ ; \ 0 \leq x \leq L\} ,$$

$$\widehat{\Gamma}_2 = \{(x, F_2(x)) \ ; \ 0 \leq x < L\}$$

such that

$$\psi(x, y) < V \text{ if and only if } F_2(x) < y < b ,$$

$$\psi(x, y) > 0 \text{ in } \Omega \text{ if and only if } 0 < y < F_1(x) .$$

We denote by

$$\Gamma_j = \{(x, f_j(x))\}$$

the portion of $\widehat{\Gamma}_j$ which lies in $\Omega$.

**Theorem 14.4** *If $\beta \in C^{m+\alpha}(\Omega) \cap C^{\alpha}(\overline{\Omega})$ for some nonnegative integer $m$, then*

*(i) $f_j(x)$ is Lipschitz continuous,*

*(ii) the curve $\Gamma_j$ is in $C^{m+1+\alpha}$, and*

*(iii) $F_j$ is continuous for $0 \leq x \leq L$.*

*Proof.* Lemma 7.3 of [6, Chap. 2] ensures that $\beta \neq 0$ on $\Gamma_j$, and then Theorem 6.1 of [6, Chap. 2] implies that $f_j(x)$ is Lipschitz continuous, and Theorem 6.2 of [6, Chap. 2] asserts the $C^{m+1+\alpha}$ regularity of the free boundary $\Gamma_j$. Finally, (iii) follows from Lemma 7.1 of [6, Chap. 2].

The assertion (ii) does not necessarily imply that function $f_j(x)$ is $C^{m+1+\alpha}$. However this is the case if $\beta_x > 0$:

**Theorem 14.5** *If in addition to (14.13), (14.16) we also assume that*

$$\frac{\partial \beta}{\partial x} > 0 \quad in \quad \Omega \, , \tag{14.17}$$

*then the curves $y = f_j(x)$ are strictly monotone increasing, and if $\beta \in C^{m+\alpha}(\Omega)$ then $f_j(x)$ is in $C^{m+1+\alpha}$*

From Theorem 1.2 of [6, Chap. 2] we also deduce the analyticity of $\Gamma_j$ (in Theorem 14.4) and of $f_j(x)$ in Theorem 14.5 provided $\beta$ is analytic.

If $\beta$ does not satisfy the condition (14.17), then the free boundaries are not necessarily monotone. For example, if the configuration in Figure 14.10 is symmetric with respect to $x = L/2$ and if $\beta(L - x, y) = \beta(x, y)$ for $0 \leq x \leq L/2$, then the solution is symmetric about $x = L/2$ and, hence, $\psi_x = 0$ at $x = L/2$. Using the maximum principle we find that $\psi_x > 0$ if $0 < x < L/2$, so that the free boundaries have the shape shown in Figure 14.11. The free boundary $\Gamma_2$ is smooth everywhere, and $f_2'(L/2) = 0$.

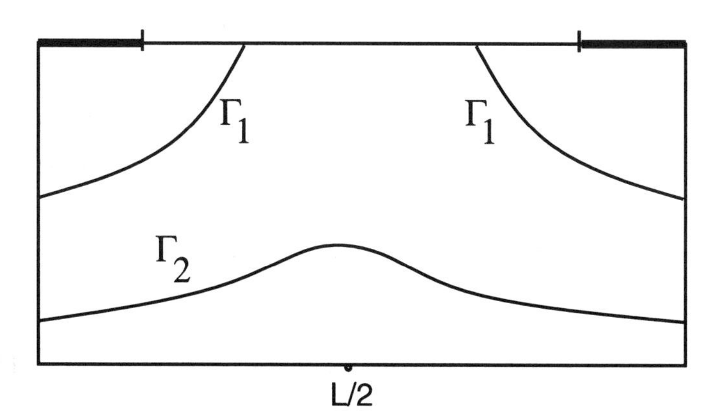


FIGURE 14.11.

In the non-symmetric configuration of Figure 14.10, there is little we can say about the shape of the free boundary. Suppose

$$\begin{aligned} \psi(x,0)=0 \quad \text{if} \quad & 0 \le x \le a \quad \text{and if} \quad L-a' \le x < L \\ & \text{where} \quad a > a' , \\ & \psi = V \quad \text{on} \quad \ell_2 , \\ & \frac{\partial \psi}{\partial n} = 0 \quad \text{elsewhere on} \quad \partial\Omega . \end{aligned} \tag{14.18}$$

Under these boundary conditions, if

$$\beta(L-x,y) \ge \beta(x,y) \quad \text{for} \quad 0 \le x \le \frac{L}{2} , \; 0 \le y \le b , \tag{14.19}$$

then

$$\psi(L-x,y) \ge \psi(x,y) \quad \text{if} \quad 0 \le x \le \frac{L}{2} , \; 0 \le y \le b . \tag{14.20}$$

Indeed, this follows by considering the difference $\psi(L-x,y)-\psi(x,y)$ (or rather its penalized approximation) and applying to it the maximum principle.

From (14.20) we deduce that

$$F_1(x-L) \ge F_1(x) , \; F_2(x-L) \ge F_2(x) \quad \text{if} \quad 0 \le x \le \frac{L}{2} . \tag{14.21}$$

In the symmetric case of Figure 14.10, with $\beta_x > 0$ if $0 < x < L/2$, the upper free boundary has precisely two components. This remains true for "nearby" $\beta$'s, by stability properties of the free boundary established in [6, Chap. 2]. However, for general $\beta$, the question of how many component the free boundary has is open.

**Theorem 14.6** *If $V$ increases then the solution $\psi = \psi_V$ increases at each point in $\Omega$.*

This follows from a general comparison result. Theorem 14.6 implies that

$$y = F_1(x)$$

is increasing with $V$, i.e., as $V$ increases $\Gamma_1$ recedes everywhere to $\{y=b\}$.

**Theorem 14.7** *If $V$ is large enough then there is no free boundary to the solution of (14.12), i.e., the depletion region is all of $\Omega$.*

To prove this result, consider the solution $\tilde\psi$ of

$$\begin{aligned} & -\Delta\tilde\psi = \beta \quad \text{in} \quad \Omega , \\ & -\tilde\psi = 0 \quad \text{on } \ell_0 , \; \tilde\psi = V \text{ on } \ell_2 , \\ & \frac{\partial\tilde\psi}{\partial n} = 0 \quad \text{elsewhere on} \quad \partial\Omega . \end{aligned} \tag{14.22}$$

We shall prove that

$$0 < \widetilde{\psi} < V \quad \text{in } \Omega \text{ if } V \text{ is large enough;} \tag{14.23}$$

it would then follow that $\widetilde{\psi}$ is the solution to the variational inequality, and it clearly has no free boundaries.

*Proof.* Write $\widetilde{\psi} = \psi_0 + V\psi$, where

$$-\Delta\psi_0 = \beta \quad \text{in} \quad \Omega ,$$

$$\psi_0 = 0 \quad \text{on} \quad \ell_0 \cup \ell_2 ,$$

$$\frac{\partial\psi_0}{\partial n} = 0 \quad \text{elsewhere on} \quad \partial\Omega ,$$

and

$$-\Delta\psi_1 = 0 \quad \text{in} \quad \Omega ,$$

$$\psi_1 = 0 \quad \text{on } \ell_0 \ , \ \psi_1 = 1 \text{ on } \ell_2 ,$$

$$\frac{\partial\psi_1}{\partial n} = 0 \quad \text{elsewhere on} \quad \partial\Omega .$$

By the maximum principle $0 < \psi_1 < 1$ and, by decreasing rearrangements, $\partial\psi_1/\partial y \leq 0$; the strong maximum principle then shows that $\partial\psi_1/\partial y < 0$ in $\Omega$. We shall first prove that

$$\widetilde{\psi} < V \quad \text{in} \quad \Omega .$$

If $0 < y < b/2$ then

$$\frac{\partial\psi_1}{\partial y} \leq -\sigma < 0 \quad (\sigma \quad \text{constant})$$

and therefore

$$\frac{\partial\widetilde{\psi}}{\partial y} = \frac{\partial\psi_0}{\partial y} + V\frac{\partial\psi_1}{\partial y} \leq C - \sigma V < 0 ,$$

so that $\widetilde{\psi} < V$. On the other hand if $(b/2) < y < b$ then, since $\partial\psi_1/\partial y < 0$,

$$\psi_1 \leq \theta < 1$$

and

$$\widetilde{\psi} = \psi_0 + V\psi_1 \leq \varphi_0 + \theta V < V .$$

Next we prove that $\widetilde{\psi} > 0$ in $\Omega$. By the maximum principle

$$\psi_1 > 0 \quad \text{on} \quad \ell_1 .$$

We also have

$$\frac{\partial \psi_1}{\partial y} < 0 \quad \text{on } \ell_0 \text{ and in all of } \Omega.$$

From this we deduce that, for some small $\varepsilon > 0$,

$$\psi_1 > c_0 r^{1/2} \cos \frac{\varphi}{2} \qquad (c_0 > 0) \tag{14.24}$$

on the portions

$$\{x = a \pm \varepsilon \ , \ b - \varepsilon \leq y \leq b\} \ ,$$

$$\{-\varepsilon \leq x - a \leq \varepsilon \ , \ y = b - \varepsilon\}$$

of the boundary of the rectangle $R_1 = \{|x - a| < \varepsilon \ , \ -\varepsilon \leq y - b \leq 0\}$. Since

$$\psi_1 = 0 \quad \text{on} \quad \ell_0 \ , \ \frac{\partial \psi_1}{\partial y} = 0 \quad \text{on} \quad \ell_1 \ ,$$

we can apply the maximum principle to deduce that (14.24) holds throughout $R_1$.

On the other hand by general regularity results at the point $Z$,

$$\psi_0 \leq C r^{1/2} \cos \frac{\varphi}{2} \quad \text{in} \quad R_1$$

and hence

$$\widetilde{\psi} = \psi_0 + V\psi_1 > 0 \quad \text{in} \quad R_1 \ .$$

Outside $R_1$ we can establish more quickly that $V\psi_1 > |\psi_0|$ , so that $\widetilde{\psi} > 0$ everywhere in $\Omega$.

From Theorem 14.7 it follows that there is a value $V = V_j$ such that the free boundary $\Gamma_j$ disappears as $V \uparrow V_j$. Let us examine how this occurs, and take for definiteness the case $j = 2$. We shall suppose for simplicity that $\beta_x \geq 0$, so that $\Gamma_2$ is given by a monotone increasing curve

$$y = f_{2,V}(x) \quad x_V < x < L \ . \tag{14.25}$$

**Theorem 14.8** *If $V \uparrow V_2$ then, generically, $x_V \to L$.*

This is illustrated in Figure 14.12

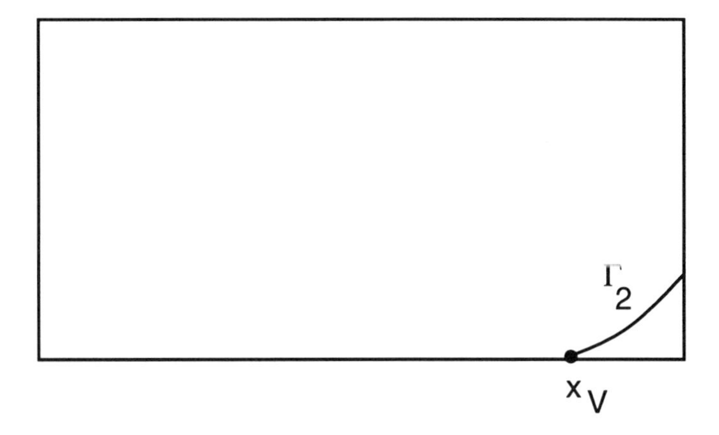


FIGURE 14.12.

*Proof.* If $x_V \to \sigma < L$ then $\widehat{\psi} = \psi_{V_2}$ satisfies

$$\widehat{\psi}(x,0) = V \quad \text{if} \quad 0 < x < L \ , \tag{14.26}$$

$$\frac{\partial \widehat{\psi}(x,0)}{\partial y} = 0 \quad \text{if} \quad \sigma < x < L \ . \tag{14.27}$$

In view of the other boundary conditions on $\widehat{\psi}$, the Cauchy data (14.26), (14.27) overdetermines $\widehat{\psi}$, in general. For example, if $\beta$ is analytic, the boundary conditions (14.26) (for $\sigma < x < L$) and (14.27) alone already determine $\widehat{\psi}$ uniquely, and so the remaining boundary conditions on $\partial\Omega \backslash \ell_2$ will not be satisfied in general. A simple example is given by

$$\widehat{\psi} = V - (y^4 + y^2)(2L - x) \ , \ \beta = -(12y^2 + 1)(2L - x) \ ;$$

here $\beta$ satisfies all the conditions required above, including $\beta_x > 0$.

All the results established in Theorems 14.1–14.8 hold also for 3-d domains, and the proofs are similar.

## 14.5 Numerical methods

There are various numerical methods for computing the free boundary for a variational inequality; see [7] for a comprehensive treatment. A quick and direct method is to use finite elements to solve the minimization problem (14.12). Since however the gradient of $\psi$ vanishes on the free boundary, one cannot expect to determine the free boundary very precisely from the $\varepsilon$-level and $(V-\varepsilon)$-level curves of the solutions. On the other hand, the above method can provide a good initial guess, after which one may continue to approximate the free boundary as follows.

Take the initial guess, say $\Gamma_{10}$ and $\Gamma_{20}$, and compute the corresponding solution $\psi_0$ of

$$-\Delta\psi_0 = \beta$$

in the domain $\Omega_0$ determine by $\Gamma_{10}$ and $\Gamma_{20}$, with boundary conditions

$$\psi_0 = 0 \text{ on } \Gamma_{10}\ ,\ \psi_0 = V \text{ on } \Gamma_{20},$$

$$\frac{\partial\psi_0}{\partial n} = 0 \quad \text{elsewhere on} \quad \partial\Omega_0\ .$$

Evaluate $\partial\psi_0/\partial n$ along $\Gamma_{10}$ and along $\Gamma_{20}$ and move these curves (see Figure 14.13) to $\Gamma_{11} = \Gamma_{10} + \varepsilon\widetilde{\Gamma}_{10}$ , $\Gamma_{21} = \Gamma_{20} + \varepsilon\widetilde{\Gamma}_{20}$ in a way that decreases $|\partial\psi_0/\partial n|$. Compute the solution $\psi_1$ to

$$-\Delta\psi_1 = \beta$$

in the new domain $\Omega_1$ determined by $\Gamma_{11}$ and $\Gamma_{21}$, and continue. In carrying out the computation of $\partial\psi_1/\partial n$ on the new boundaries $\Gamma_{11}, \Gamma_{21}$ we can use Hadamard's formula which describes how Green's function for $\Omega_1$ differs from Green's function for $\Omega_0$, to the order $\varepsilon$; see, for example [8, Chapter 1]. The fact (established in §14.4) that the free boundary is smooth suggests (by general stability results [6, Chap. 2, §10]) that if we make a good initial guess of $\Gamma_{10}, \Gamma_{20}$ then the above method will converge. However, if we make a bad initial guess, the above approach for moving the boundary will in general develop instabilities. This is due to the fact that the Cauchy problem for the elliptic equation $-\Delta\psi = \beta$ is ill-posed.

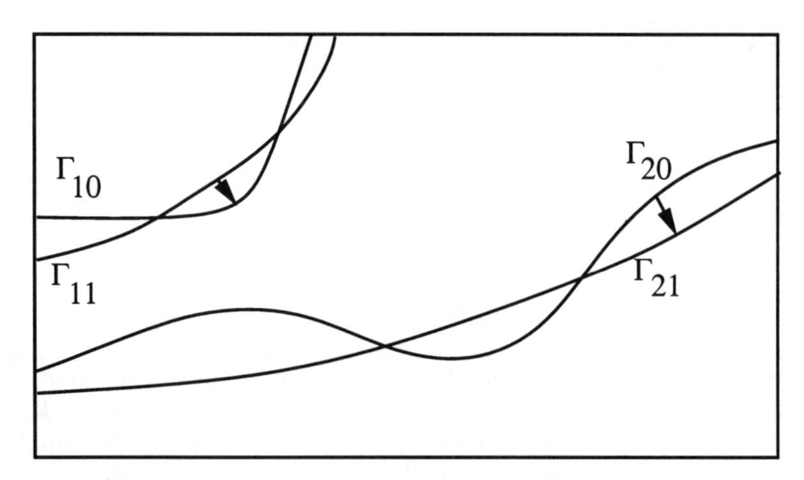


FIGURE 14.13.

## 14.6 References

[1] A. Friedman, *Mathematics in Industrial Problems, Part 7*, IMA Volume 67, Springer–Verlag, New York (1994).

[2] A. Friedman, *Mathematics in Industrial Problems, Part 2*, IMA Volume 24, Springer–Verlag, New York (1989).

[3] S. Selberherr, *Analysis and Simulation of Semiconductor Devices*, Springer–Verlag, New York (1986).

[4] A.S. Grove, *Physics and Technology of Semiconductor Devices*, John Wiley and Sons, New York (1967).

[5] A.S. Grove, *Physics and Technology of Semiconductor Devices*, John Wiley and Sons, New York (1967).

[6] A. Friedman, *Variational Inequalities and Free-Boundary Problems*, Wiley–Interscience, New York (1982).

[7] R. Glowinski, J.-L. Lions and R. Tremolières, *Numerical Analysis of Variational Inequalities*, North-Holland, Amsterdam (1981).

[8] Z. Nehari, *Conformal Mapping*, McGraw–Hill, New York (1952).

# 15

# Causal modeling in diagnosing complex industrial equipment

Real-time monitoring and diagnostic systems play a critical role in maintaining performance and improving service of complex industrial products such as utility turbine-generators, jet engines, medical equipment, locomotives, etc. The systems detect early indications of equipment failures, identify the underlying causes and provide advice on corrective actions. A key challenge in the development of effective systems is the representation of the complex diagnostic knowledge and its associated reasoning mechanism. On April 26, 1996 Mahesh Morjaria from General Electric Corporate R&D described a causal modeling technique that meets these challenges and offers significant advantages over other approaches. The computational issues however need to be advanced in order to make this technique applicable to future more complex industrial equipment.

## 15.1 Monitoring and diagnostic systems

We are concerned here with industrial equipment that has the following characteristics:

(i) Complex design, and therefore difficult to diagnose;

(ii) High reliability and availability, with many diverse but infrequent failure modes;

(iii) Highly customized equipment, i.e., various design options and differing operating practices;

(iv) Cost of forced outage is very high;

(v) Typically long product life (average more than 20 years).

Examples for such *GE* products include utility turbine-generators, locomotives, jet engines, and medical equipment.

Each of these complex products has on-site monitoring and diagnostic system whose major functions are to:

(a) detect early indications of equipment failures,

(b) identify underlying causes, and

(c) provide advice on corrective actions.

The customer/business drivers are: improving equipment performance/ maintenance/availability, improving service quality, and reducing service delivery cost.

A rule-based expert system which utilizes a set of diagnostics rules is typically used for implementing diagnosis. However, in diagnosing complex equipment, such as those mentioned above, the cause and effect relationships are only probabilistic, and more complex modeling is needed than the simple rule-based system.

## 15.2 Causal model

A *causal model* represents cause and effect relationships between potential equipment failures and indicators for those failures. The model is developed by experts who are intimately familiar with the various functions of the equipment and potential failure modes. A well defined set of semantics have been developed to represent various types of causal relationships. The model also includes expert assessment of the strength of each causal relationship and the likelihood of root cause failure. A set of indicators that provide diagnostic evidence to the model are also identified.

The causal model is then transformed into a well formulated Bayesian belief network (to be precisely defined in the next section). The network is a directed acyclic graph in which nodes represent probabilistic variables and the edges indicate conditional dependence of one variable on another. The transformation from a causal model to a belief network depends upon the types of relationships present in the model. The belief network provides formal means to probabilistic inferencing, that is, when information about a variable's state is given to the network in the form of evidence, an update algorithm computes the probability (or belief) distribution of the other variables. The updated beliefs are then used to infer the beliefs of the failures in the causal model.

Figure 15.1 taken from [1] (some of the work was reported in [2]) is an example of a portion of Bayesian belief network. The nodes represent random variables of interest and their states (e.g., Hydrogen Cooler Leak; True/False); there is directed relationship between variables, probability is associated with state of each variable (*belief*), a priori probability numbers are associated with all the root nodes, and conditional probability numbers are associated with all the links.

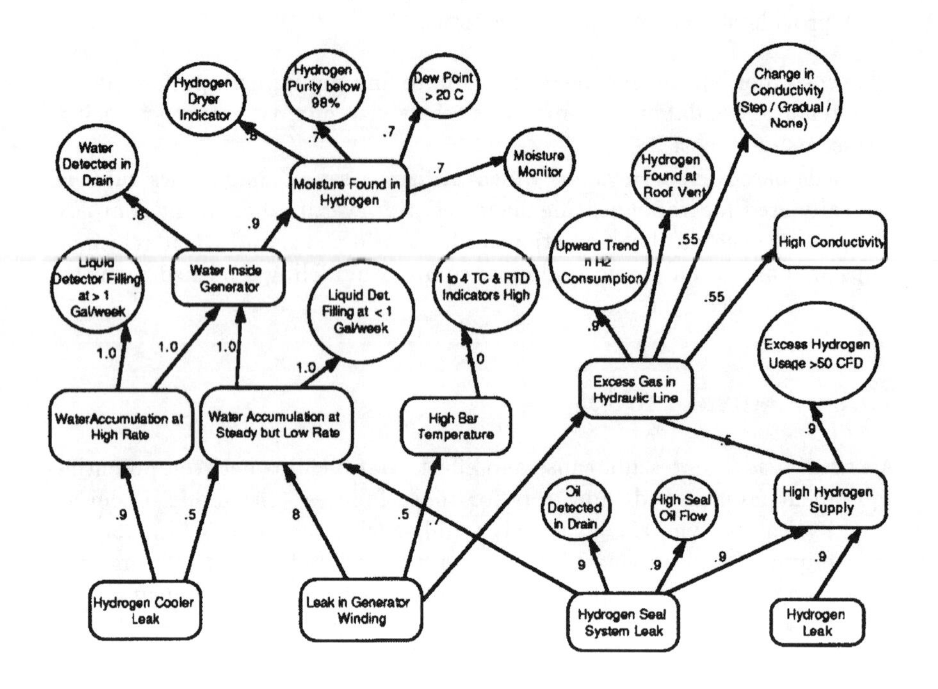


FIGURE 15.1.

In the next sections we shall describe a mechanism for propagating the influence by available evidence to other nodes, thereby updating beliefs.

## 15.3 Causal (belief) network

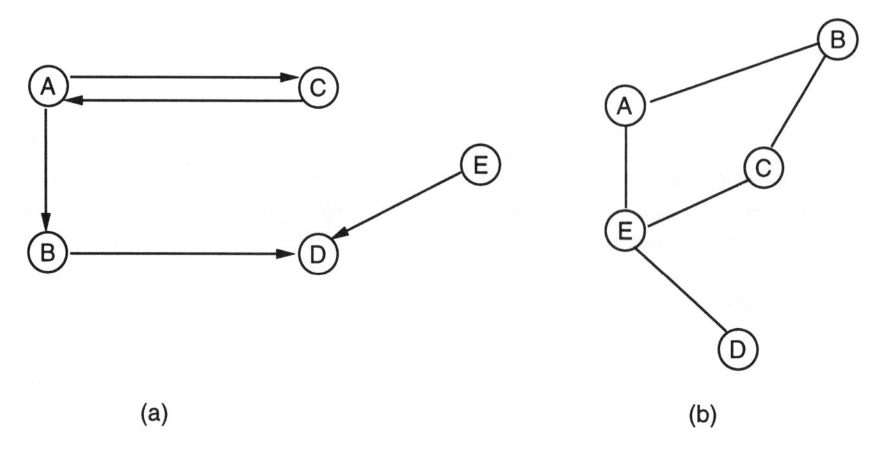


FIGURE 15.2.

A graph $G$ consists of a finite set $V$ of vertices, or nodes, and a subset $E$ of pairs of distinct vertices (called edges, or links). The graph is *directed* if the pairs are ordered, and *undirected* if the pairs are not ordered. Figure 15.2(a) and (b) show examples of directed and undirected graphs, respectively.

A *path* $\{v_0, v_1, \ldots, v_n\}$ is a sequence of vertices $v_i$ such that each pair $(v_i, v_{i+1})$ is in $E$. A path from a vertex $v$ to itself is called a *cycle*. Undirected graph is *complete* if every pair of distinct vertices is an edge. In a graph $G = (V, E)$, every subset $W$ of $V$ induces a subgraph $G_W$: its edges are all the edges in $E$ with endpoints in $W$.

If $G = (V, E)$ is undirected graph then a subset $W$ of $V$ is called a *complete* set if it induces a complete graph. A subset $W$ of $V$ is called a *clique* if it is maximal complete set.

An undirected graph is called *triangulated* if any simple cycle (i.e., cycle with no repeating vertices) of length $> 3$ has a chord, i.e., an edge connecting two non-adjacent vertices of the cycle. Figure 15.3(a) shows a triangulated graph, and Figure 15.3(b) shows untriangulated graph (the cycle ABEC has no chord).

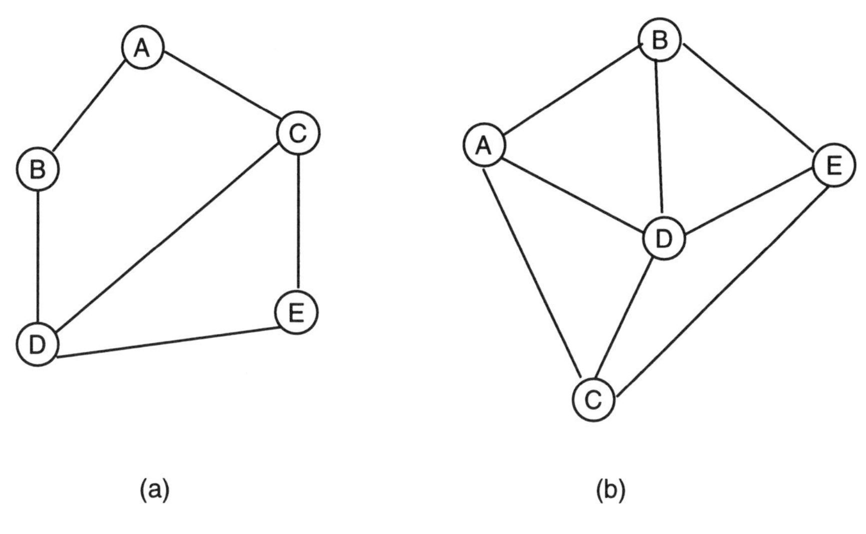


FIGURE 15.3.

*Acyclic* graph is a graph that has no cycles. If a graph is directed and acyclic then we call it *directed acyclic graph* (DAG).

Let $G = (V, E)$ be a directed graph and let $u, v$ be vertices in $V$. If $(u, v) \in E$ then $u$ is called a *parent* (predecessor) of $v$ and $v$ is called a *child* (successor) of $u$. We say that $u$ is an *ancestor* of $v$ and $v$ is a *descendent* of $u$ if there is a path from $u$ to $v$. A vertex with no parent is called a *root*; a vertex with no child is called a *leaf*.

A DAG $(V, E)$ is *singly connected* if for every $u$ and $v$ in $V$ there is at most one chain between $u$ and $v$. A DAG is called a *forest* if every vertex

$v$ has at most one parent; a forest is clearly singly connected. A forest is called a *tree* if exactly one vertex has no parent; this vertex is called the *root* of the tree.

Following [3] we shall now proceed to define causal or (belief) network. Each vertex will be a proposition, and cause-and-effect between propositions will be described by directed edges with assigned probabilities. The cause-and-effect requirement dictates that there should be no cycles.

Let $(\Omega, \mathcal{F}, P)$ be a given probability space. A *propositional variable* $A$ on the probability space is a function from $\Omega$ to a space whose elements $E_1, \ldots, E_n$ are sets of $\mathcal{F}$ which are exclusive and exhaustive, i.e.,

$$E_i \cap E_j \neq 0 \quad \text{if} \quad i \neq j \,, \qquad \bigcup_{i=1}^{n} E_i = \Omega \,.$$

If $E_i$ corresponds to a statement, say $A = a_i$, then we write $P(E_i) = P(A = a_i) = P(a_i)$. If a propositional variable has been assigned a known value, say $a_i$, then we say that $A$ has been *instantiated* at $a_i$.

Let $B$ be another propositional variable with exclusive and exhaustive events $F_i$ $(1 \leq i \leq m)$, so that $F_i = (B = b_i)$. We define intersection

$$P(a_i \wedge b_j) = P(E_i \cap F_j)$$

and conditional probability

$$P(a_i|b_j) = P(E_i|F_j) = \frac{P(E_i \cap F_j)}{P(F_j)} = \frac{P(a_i \wedge b_j)}{P(b_j)}$$

if $P(b_j) > 0$. By Bayes' Theorem

$$P(F_i|A) = \frac{P(F_i)P(A|F_i)}{\sum_{k=1}^{m} P(F_k)P(A|F_k)}$$

for any $A$ in $\mathcal{F}$ and, in particular,

$$P(b_i|a_j) = \frac{P(b_i)P(a_j|b_i)}{\sum_{k=1}^{m} P(b_k)P(a_j|b_k)}$$

An *inference network* is a directed acyclic graph (DAG) in which the set of vertices is a set of propositions.

*Definition 15.1.* Let $V$ be a finite set of propositional variables on a probability space $(\Omega, F, P)$, and let $G = (V, E)$ be a DAG. For each $v \in V$ let $c(v)$ denote the subset of all parents of $v$, and $a(v)$ the subset of all variables in $V$ excluding $v$ and the set $d(v)$ of all its descendents. Suppose that, for any subset $W$ in $a(v)$, $W$ and $v$ are conditionally independent given $c(v)$, that is, if $P(c(v)) > 0$ then either

$$P(v|c(v)) = 0 \quad \text{or} \quad P(W|c(v)) = 0 \quad \text{or} \quad P(v|W \cup c(v)) = P(v|c(v)) \,. \tag{15.1}$$

Then $C = (V, E, P)$ is called *causal network*, and $c(v)$ is the set of *causes* of $v$.

The concept of conditional probability $P(X|Y)$ used in (15.1) is defined as follows: If $X$ and $Y$ are random variables with finite range $\{x_i\}$ and $\{y_j\}$ respectively, then $P(X|Y)$ is the vector with components

$$P(X = x_i | Y = y_j) = \frac{P(X = x_i,\ Y = y_j)}{P(Y = y_j)} \,.$$

Consider a causal network such as in Figure 15.4 (taken from [3, p. 218]). (Note that from the information provided in Figure 15.4 we can compute the probabilities of all the propositional variables, and also the conditional probabilities from child to parent (by Bayes' Theorem).) We are interested in questions like: If $A$ and $C$ are instantiated at $a_1$ and $c_2$, what is the probability of $B = b_1$, or of $D = d_1$, etc. ?

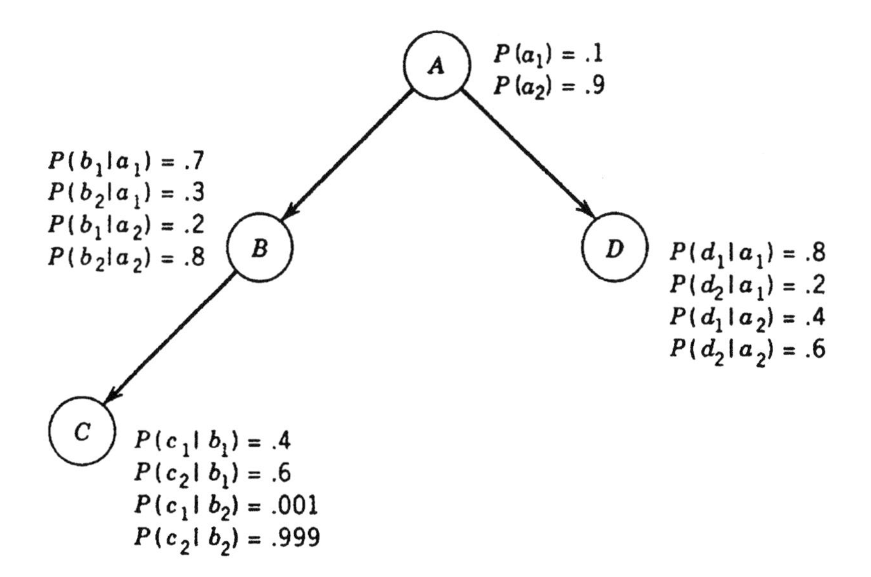


FIGURE 15.4.

J. Pearl [4] (see also [3, Chap. 6]) developed a method that systematically computes the probability of events in singly connected graphs; this is known as the method of *probability propagation*. We first describe the method for causal network which are trees.

Let $W$ be a subset of variables which are instantiated, i.e., specific values are assigned to them. For any $B \in V$ let $W_B^-$ = subset of $W$ in the tree rooted at $B$, $W_B^+ = W - W_B^-$. Then [3, p. 209]

$$P(b_i|W) = \beta_B P(W_B^-|b_i)P(b_i|W_B^+) \tag{15.2}$$

where $b_i$ are the values of $B$ and $\beta_B$ is a normalizing factor (depending on $B$ and $W$).

Define

$$\lambda(b_i) = \begin{cases} P(W_B^-|b_i) \quad \text{if} \quad B \notin W \\ 1 \quad \text{if} \quad B \in W \text{ and } b_i \text{ is the instantiated value} \\ 0 \quad \text{if } B \in W \text{ and } b_i \text{ is not the instantiated value,} \end{cases} \tag{15.3}$$

$$\pi(b_i) = P(b_i|W_B^+) \,. \tag{15.4}$$

If we can compute the $\lambda(b_i)$, $\pi(b_i)$, then we can also find (from (15.2)) the desired probabilities $P(b_i|W)$.

To compute $\lambda(b_i)$, $\pi(b_i)$ we introduce for any $B \in V$ and $C$ child of $B$

$$\lambda_C(b_i) = \sum_{j=1}^{m} P(c_j|b_i)\lambda(c_i) \tag{15.5}$$

where $C$ has events $c_1, \ldots, c_m$. The vector $\lambda_C(B)$ with components $\lambda_C(b_i)$ is called $\lambda$ *message* from $C$ to $B$ (child to parent). If $\lambda(C)$ (the vector with components $\lambda(c_i)$) is known, then we can immediately compute $\lambda_C(B)$ since the $P(c_j|b_i)$ are given numbers.

Note that if $B$ is a leaf then $\lambda(b_i) = 1$, and if $B$ is a root then $\pi(b_i) = P(b_i)$.

Next, for any $B$ in $V$ which is not a root and $A$ parent of $B$, we define

$$\pi_B(a_j) = \begin{cases} 1 \text{ if } A \text{ is instantiated at } a_j \\ 0 \text{ if } A \text{ is instantiated, but not at } a_j \\ \pi(a_j) \prod\limits_{\substack{C \in s(A) \\ C \neq B}} \lambda_C(a_j) \text{ if } A \text{ is not instantiated} \end{cases} \tag{15.6}$$

where $s(A)$ is the subset of all children of $A$. The vector $\pi_B(A)$ with components $\pi_B(a_j)$ is called the *message* $\pi$ from $A$ to $B$ (from parent to child).

The following two theorems allow us to compute the $\lambda$ messages step-by-step from leaves all the way to root and $\pi$ messages from root all the way to leaves:

**Theorem 15.1** *[3, p. 213].*

$$\lambda(b_i) = \begin{cases} \rho_B \prod_{C \in s(A)} \lambda_C(b_i) & \text{if } B \notin W \\ 1 \text{ if } B \in W \text{ and } B \text{ is instantiated at } b_i \\ 0 \text{ if } B \in W \text{ and } B \text{ is not instantiated at } b_i \; . \end{cases} \tag{15.7}$$

**Theorem 15.2** *[1, p. 214]*

$$\pi(b_i) = \mu_B \sum_{j=1}^{m} P(b_i|a_j)\pi_B(a_j) \; . \tag{15.8}$$

Here $\rho_B$ and $\mu_B$ are normalizing constants. If we substitute $\lambda_C$ from (15.5) into (15.7), we get iterative formula which propagates the $\lambda$ message from child to parent. If we substitute $\pi_B$ from (15.6) into (15.8) we get iterative formula which propagates the $\pi$ message from parent to child.

Consider the example in Figure 15.4. It can serve as a simple model of spouse cheating [3, pp. 161, 217] where

$a_1$ = spouse is cheating,

$c_1$ = spouse dines with another,

$b_1$ = spouse is reported seen dining with another,

$d_1$ = strange person calls on the phone

and $a_2, b_2, c_2, d_2$ are the negations of the above events. If $W = \emptyset$ then $P(a_1) = 0.1$ whereas if $B$ is instantiated for $b_1$ then the probability of spouse cheating increases to $P(a_1|W) = 0.28$. This last result can be obtained by using the above methodology in a straightforward way.

The above methodology can be easily extended to singly connected DAG [3, pp. 237–240]. If the graph is not singly connected, it can be broken into singly connected DAG by *conditioning* or *cutset*. We illustrate this in the case of the directed graph in Figure 15.5 (cf. [3, p. 248]).

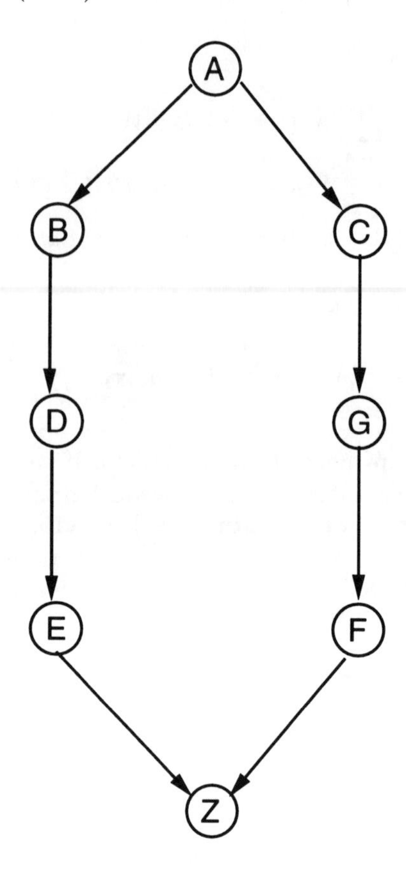


FIGURE 15.5.

We want to compute $P(Z|g_2)$, where $G$ is instantiated at $g_2$. We "condition" (or "cut") $A$ by taking $a_1$ with conditional probability $P(a_1|g_2)$ and $a_2$ with conditional probability $P(a_2|g_2)$. We then compute $P(Z|g_2, a_1)$, $P(Z|g_2, a_2)$ by the previous method for singly connected graphs ($A$ is removed from the graph). Finally we compute $P(Z|g_2)$ from the formula

$$P(Z|g_2) = P(Z|g_2, a_1)P(a_1, g_2) + P(Z|g_2, a_2)P(a_2|g_2) \; .$$

Complexity in the cutset method is measured in terms of the number of the cutset nodes (In the preceding example the cutset node consists of just the node $A$.) Computer time to solve a network is exponential in number of cutset nodes. In actual complex equipment monitoring the number of cutset points is large, and so other methods for solving the network need to be developed.

## 15.4 Causal network with cliques

One such method was devised by Lauritzen and Spieghalter [5]. It is based on the idea of viewing any causal network as embedded in a causal network which is a tree, and whose vertices are cliques formed from the original graph.

*Definition 15.2.* Let $G = (V, E)$ be a DAG. If $v \in V$ and $u, w$ are parents of $v$, we form an edge between $u$ and $w$ (i.e., "marry" the parents). We do this for any $v \in V$ and add all the new edges to $E$, to obtain a larger set $E'$ of edges. The undirected graph $G_m = (V, E')$ is called the *moral graph* relative to $G$.

Now take a causal network $C = (V, E, P)$ where $G = (V, E)$ is DAG, form the moral graph $G_m$ and then extend it further (by adding edges) to obtain a triangulated graph $G_u$. Let $C\ell q_i \quad (1 \leq i \leq p)$ be the cliques of $G_u$. To each $v \in V$ we assign a unique clique $C\ell q_i$ such that

$$\{v\} \cup c\{v\} \subset C\ell q_i \ ;$$

this is possible since parents in the original graph are married and therefore $\{v\} \cup c(v)$ is a complete set in $G_m$ and then also in $G_u$, and a complete set has at least one clique. We assign to each $v$ a clique $f(v)$ (if there are several such cliques we assign just one clique, in an arbitrary way). Next we define $\psi$ by

$$\psi(C\ell q_i) = \prod_{f(v)=C\ell q_i} P(v|c(v)) \tag{15.9}$$

where the equality means that for any values of the propositional variables we have equality. It can be proved that

$$P(V) = K \prod_{i=1}^{p} \psi(C\ell q_i) \qquad (K \quad \text{constant}) \tag{15.10}$$

in the sense that, for any values of the propositional variables, equality holds. It is easy to see that the $C\ell q_i$ form a causal network with DAG which is a tree, and so the algorithm for propagating probabilities in arbitrary causal network for a tree can be applied to compute $\psi(C\ell q_i)$ [2, §7.3]. We can then use (15.10) to compute the probabilities of events in $V$, which is of course our goal.

The complexity of the method is measured in terms of the clique size; the amount of computer storage grows exponentially with clique size: Clique size of 20 requires several megabytes of storage; clique of size 40 requires processing time more than the time since the beginnning of the universe. Thus the problem of determining the "optimal" clique tree is important, yet this problem is NP-hard in terms of the number of variables in the network. Cooper [6] has shown that, for a set of nonsingly connected causal networks which are restricted to having no more than 3 parents and 5 children per variable, the problem of determining the probabilities of the remaining variables given that certain variables have been instantiated is NP-hard in terms of the number of variables in the network.

## 15.5 Challenges

Causal model-based diagnostic technology is intuitive and precise method of reasoning with diagnostic causal problems. It utilizes strong theoretical foundation of belief networks, enables sharper diagnosis by "explaining away" of likely causes, by exploiting "strong" combination of evidence, and by bi-directional belief propagation. It overcomes limitations of rule-based system which does not have means for handling uncertainty.

But the causal-based diagnostic method has limitations:

(1) Probability propagation for nonsingly network is $NP$-hard in terms of the number of variables in the network.

(2) Determining the optimal clique tree is $NP$-hard in terms of the number of variables in the network.

Expert assessment which is the basis for assigning probabilities the links of a causal network is not precise, whereas the calculations in causal network with assigned probabilities are being performed with high precision. The challenge is then to develop less precise quantitative results for the causal network calculations, but which can be executed in reduced time. Some work in this direction was done by M. Morjaria.

We finally refer to the recent monograph [7] of Shafer: it describes additional methods for computing propagation of probabilities in belief networks, and includes very recent references.

## 15.6 References

[1] M. Morjaria, M.K. Simmons and J.P. Stillman, *Development of Belief Networks Technology for Diagnostic Systems*, GE CR&D Report 92CRD081, April 1992.

[2] M. Morjaria, F.J. Rink, G. Klempner, C. Burns and J. Stein, *Commercialization of EPRI's Generator Expert Monitoring System*, Expert System Applications for the Electric Power Industry: International Conference and Exhibition, Phoenix, AZ, Dec. 1993.

[3] R.E. Neaopolitan, *Probabilistic Reasoning in Expert Systems*, Wiley-Interscience, New York (1990).

[4] J. Pearl, *Fusion, propagation and structuring in belief networks*, Artificial Intelligence, 29 (1986), 241–288.

[5] S.L. Lauritzen and D.J. Spiegelhalter, *Local computation with probabilities in graphical structures and their applications to expert systems*, Journal of the Royal Statistical Society B, 50 (1988), 157–224.

[6] G.F. Cooper, *Probabilistic inference using belief networks in $NP$-hard*, Technical Report KSL–87–27, Stanford University, Stanford California (1988).

[7] G. Shafer, *Probabilistic Expert Systems*, CBMS Regional Conference held at the University of North Dakota at Grand Forks, June 1–5, 1992, SIAM, Philadelphia (1996).

# 16

# Mathematical problems in electrical well logging

Electrical logging, which is the measurement of the (DC) electrical conductivity of rocks along the track of a well drilled into the earth, was invented in the 1920's by two French brothers, Conrad and Marcel Schlumberger, with help from Henri Doll, a French engineer. Since then electrical logging (combined with other logging measurements, mainly nuclear and acoustic) has become the chief way of quantifying the amount of oil and gas in hydrocarbon reservoirs. Well logging ranks second, behind seismic exploration, in worldwide commercial geophysical activity.

Electrical impedance is the total opposition offered by an electric circuit to the flow of electric current; it is measured in ohms. The method of electrical impedance, or applied potential, tomography is the method of mapping the electrical conductivity of a body by electrical measurements on its surface: Potential is applied at varying points by electrodes on the surface and the resulting current is measured on other electrodes. In electrical logging the potential is applied inside the body (the earth) and current is measured between points also inside the body. Thus electrical logging can be cast as an "inside-out" version of electrical impedance tomography. Practical solutions of the inverse problem have to deal with constraints on the size and number of electrodes that are deployed in a downhole array and with large contact impedance effects. On May 10, 1996, Michael L. Oristaglio from Schlumberger-Doll Research (Ridgefield, CT) described present technology, and then went on to outline a new approach based on a mathematical model of the physical problem. He posed several mathematical questions whose solution will help in developing a full 3D reconstruction of electrical conductivity near boreholes.

## 16.1 Basic facts

Oil reservoirs are accumulations of oil (and often natural gas) in the pores of rocks underground. Figure 16.1 illustrates a salt dome—a typical structure that traps oil in the (Persian and Mexican) Gulf Coast and North Sea—and a well drilled to the oil reservoirs (the dark regions). Outside the dome, the earth is layered: each layer is nearly (but not quite) horizontal. The oil is trapped by the impermeable salt region and an impermeable layer above the reservoir.

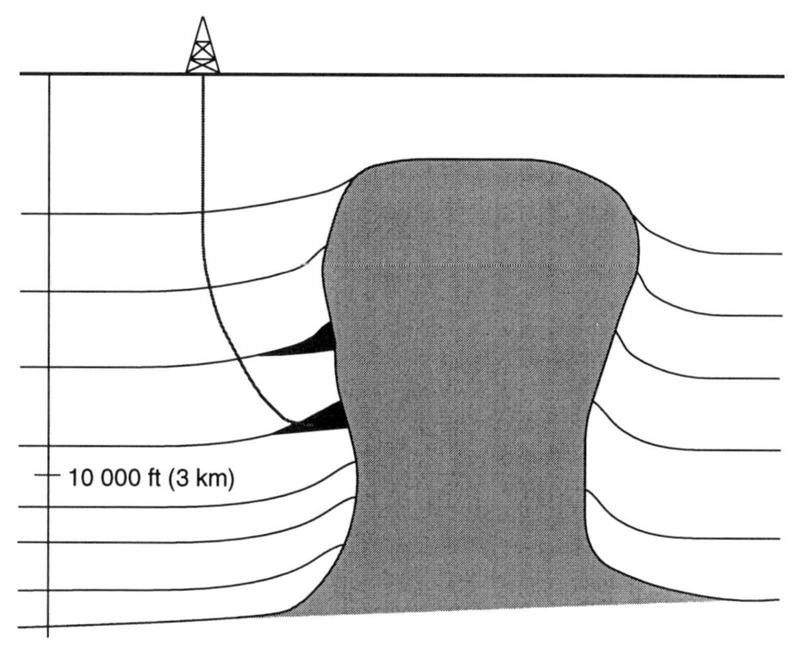


FIGURE 16.1.

Drilling cost for exploratory wells is about \$1000/ft; a deep well can easily cost \$10 million. The number of wells drilled presently in the U.S. is about 20000/year and 40000/year worldwide. Success ratio is 1 to 5 (or 10) for new territories, but is higher nearby existing fields. Figure 16.2 shows how to measure the resistance of a rock specimen. Potential $V$ is applied in a circuit which passes through the specimen, and the current $i$ is measured. By Ohm's law,

$$i = V/r$$

where $r$ is the electrical resistance. The resistivity $R$ is defined by

$$r = R\frac{L}{A},$$

where $L$ is the length of the specimen and $A$ is its cross-sectional area; the electrical conductivity is defined by

$$\sigma = \frac{1}{R}.$$

$R$ is measued in ohm-meter ($\Omega\,\mathrm{m}$) and $\sigma$ is measured in siemens/meter (S/m); siemens = 1/ohm.

Rock is a porous material. The pores in rock layers contain fluid, mostly (salt) water. In oil reservoirs the pores contain also oil to varying degrees. This affects the conductivity of the rock, because oil has (effectively) zero

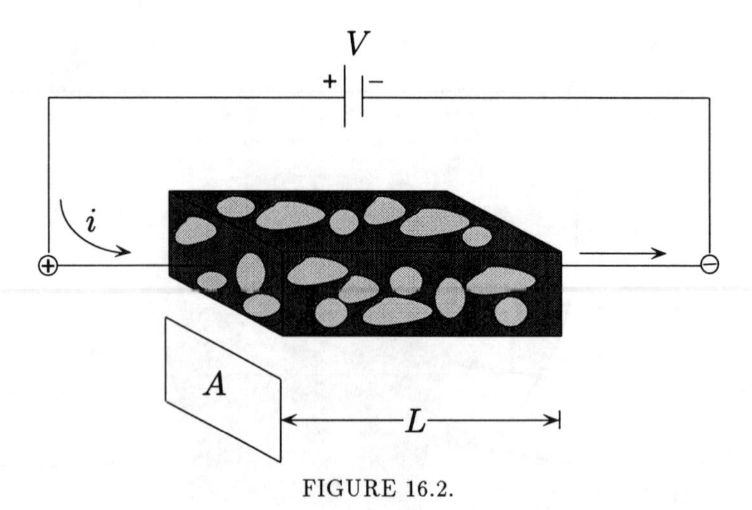


FIGURE 16.2.

conductivity. Thus from electrical logging one can infer, with some level of confidence, the presence (or absence) of oil in rock layers. Nowadays electrical logs can be recorded while drilling and used to change the direction of the drilling (as sketched in Figure 16.1) toward an oil reservoir.

Let $\phi$ denote the porosity of rock; i.e., the pores in volume $V$ occupy volume $\phi V$. Let $S_w$ and $S_{\text{oil}}$ denote the fractional volumes of pores occupied by water and oil, respectively; $S_{\text{oil}} = 1 - S_w$. Denote by $\sigma_w$ and $R_w$ the conductivity and resistivity of water, and by $\sigma_t$ and $R_t$, the conductivity and resistivity of the mixture—oil, water, and rock. Let the rock matrix (solid part) have zero conductivity. Archie's law [1] is a practical rule which relates the amount of oil in the rock to the conductivity or resistivity of the mixture:

$$\sigma_t = \sigma_w (S_w \phi)^2$$

or

$$S_{\text{oil}} = 1 - \frac{1}{\phi}\left(\frac{R_w}{R_t}\right)^{1/2}.$$

## 16.2 Azimuthal resistivity imager*

Figure 16.3 shows a profile of borehole with boundary $\Gamma$. A pump circulates mud through the borehole during drilling to lubricate the drilling bit and flush rock cuttings to the surface. Some of the mud leaks into surrounding permeable rocks creating an altered ("invaded") zone near the borehole. Outside the borehole the various rock layers are nearly parallel, with sharp

*Mark of Schlumberger

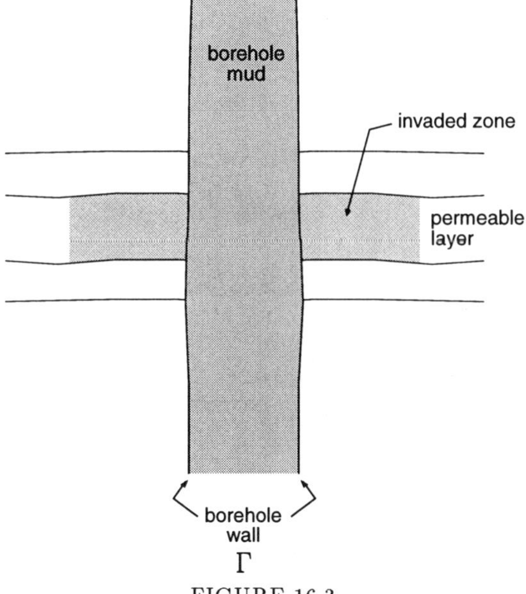


FIGURE 16.3.

transition widths: The layers can be several meters thick, whereas the transition width is a cm or less. The conductivities of the layers range between about 0.001 S/m (resistivity of 1000 $\Omega$ m) to 2 S/m. The conductivity of some drilling muds ("water-based") is much higher, up to 40 S/m. The invaded zone can extend out to a radius of 60 inches from the borehole. Figure 16.4 shows a possible profile of the conductivity going radially away from the borehole axis through the invaded zone in a permeable layer. The bump in the conductivity in the middle of the profile (called an "annulus") results from the interaction of two diffusive processes: hydraulic diffusion

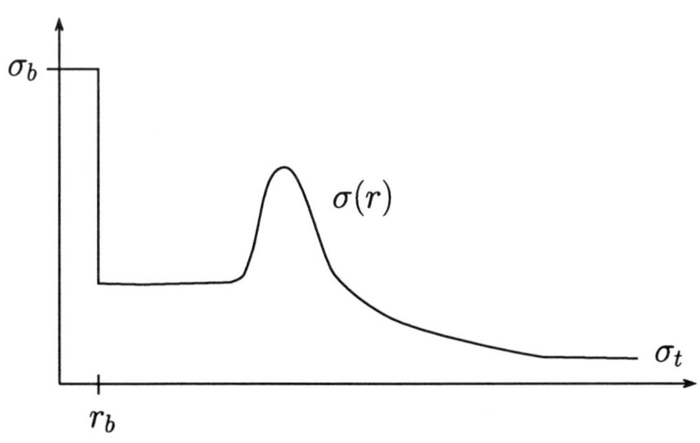


FIGURE 16.4.

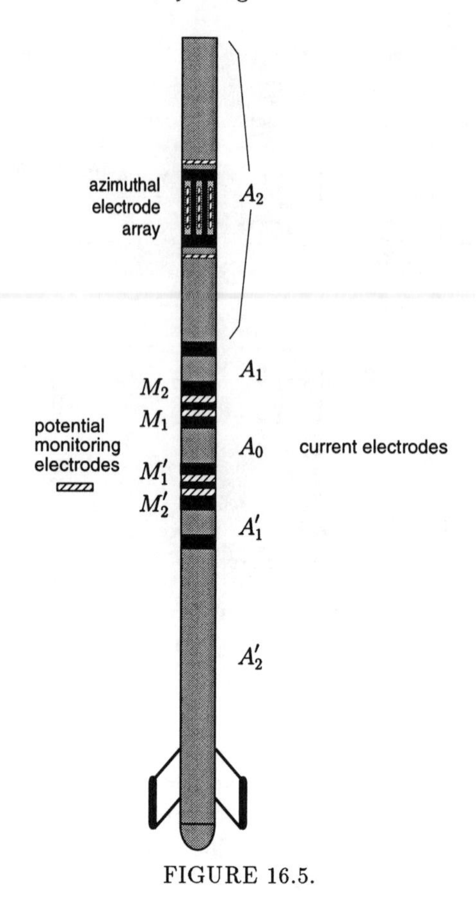


FIGURE 16.5.

of the mud outward from the borehole through the permeable layer and diffusion of dissolved salts in the fluid from higher concentration (in the borehole mud) to lower concentration (in the native formation fluids).

Figure 16.5 shows schematically an instrument developed and used by Schlumberger for electrical logging; it is called Azimuthal Resistivity Imager or ARI* [2]. Its length is 33.7 ft and its diameter is 3-5/8 inches. The ARI consists of a modern dual laterolog tool [3] and an array of azimuthal electrodes. In the dual laterolog system, which is the present standard for resistivity logging, electrodes $A_0$, $A_1$, $A_2$, $A_1'$, $A_2'$ (in grey) inject current into the earth while electrodes $M_1$, $M_2$, $M_1'$, $M_2'$ monitor the potential along the sonde (the dark bands are insulating sections) ["Sonde" refers to any device for measuring and telemetring physical conditions.]. The potential on the $A$ electrodes is also measured; the $M$ electrodes, however, are connected to high-impedance electronics so that (almost) no current flows through them. A cable supplying power and communication with a surface computer lowers the tool into the borehole; measurements are (usually) made continously as the tool is pulled up from the bottom of the hole. The potentials are measured with respect to a grounded electrode separated from the tool by a special insulated section of the cable.

Current is sent between two $A$ electrodes or between $A$ electrode and a surface electrode, and then the voltage is measured on all the electrodes. Assuming that the transformation from current data $I_j$ to corresponding potential data is linear, the linear transfer (impedance) matrix is then computed, and it is used to determine the (average) conductivity of the layer at the given depth. The dual laterolog actually operates in two separate modes: a shallow mode in which all current emitted from the central $A_0$ electrode returns on the auxiliary $A$ electrodes, and a deep mode in which net current is emitted from all the $A$ electrodes and returns to the surface electrode. The measurements are most sensitive to the conductivity in the layer(s) of the earth at the depths between the monitoring electrodes, $M_2$ and $M_2'$. The shallow and deep measurements are combined to estimate the (average) conductivity in the invaded and uninvaded zones. The (12) azimuthal electrodes on the ARI allow a third mode of operation that gives information on the variation of conductivity around the borehole. For more details, see [2], [3], and [4] and the references therein.

## 16.3 Mathematical model

In this section we shall describe a mathematical model based on the physics of the problem. The goal is to use the model to obtain more precise computational results for electrical logging than those obtained with the dual laterolog described at the end of §16.2. Figure 16.6 shows a model of the earth symmetric about an axis of symmetry along the center of a borehole. Coordinates $\mathbf{r} = (\mathbf{r}, \mathbf{z})$ measure radial distance $r$ from the axis and depth $z$ along the axis (increasing downwards). The radius of the borehole $r_b(z)$ is variable. The central part of the borehole out to a radius $r = r_s$ is occupied by a measuring apparatus ("tool"), which includes a sequence of electrodes $j = 1, 2, \ldots, J$, covering a certain depth interval. We denote by $S$ the lateral surface of the tool, and by $\Omega$ the domain outside the tool. The lateral surface of the $j$-th electrode is given by

$$S_j : r = r_s, \; z_j^{(b)} < z - z_s < z_j^{(t)} ,$$

where $z_s$ is the depth of a reference point (e.g., the center) of the electrode array. Let $\chi_j(\mathbf{r})$ denote the function on $S$, satisfying

$$\chi_j = 1 \quad \text{on} \quad S_j, \quad \chi_j = 0 \quad \text{on} \quad S \backslash S_j .$$

We take (typical values) $r_s = 4.6$ cm, $L = 8$ m (length of electrode array), the average of $r_b = 10$ cm, $\delta r_b \approx 2$ cm (variation of $r_b$), the conductivity $\sigma$ outside the borehole between 0.001 and 1 S/m, and $\sigma_b$, the conductivity

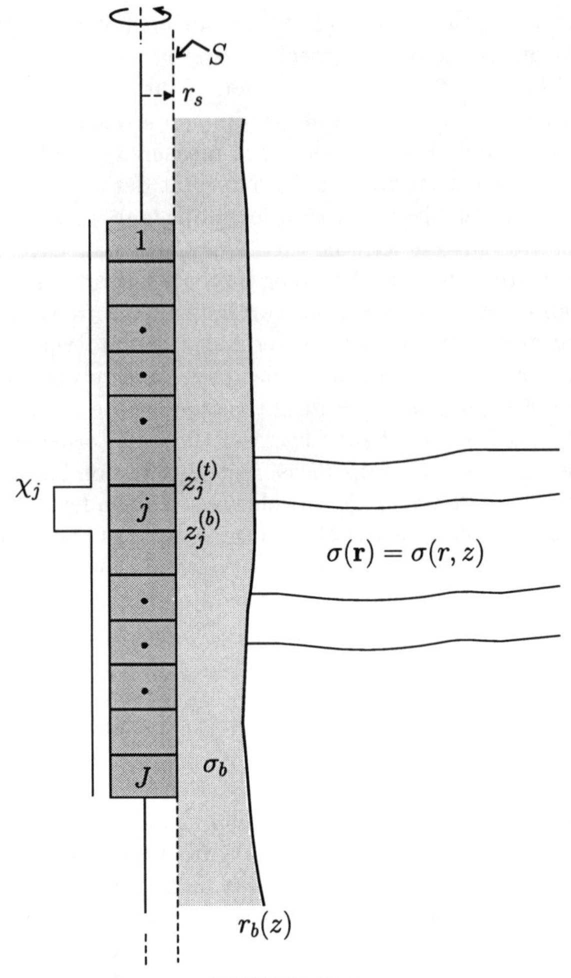


FIGURE 16.6.

inside the borehole to be between 10 and 40 S/m. The array is moved in increments $\delta z = 15$ cm along the borehole axis.

At each position of the array, we apply potential $V_j$ in turn to each electrode $j$, and calculate the current on electrode $i$. The potential $\psi_j$ satisfies the differential equation

$$\nabla \cdot (\sigma(\mathbf{r}) \nabla \psi_{\mathbf{j}}(\mathbf{r})) = \mathbf{0} \quad \text{in} \quad \mathbf{\Omega}, \tag{16.1}$$

and the boundary condition

$$\psi_j(\mathbf{r}) = \mathbf{V_j} \chi_{\mathbf{j}}(\mathbf{r}) \quad \text{on} \quad \mathbf{S}. \tag{16.2}$$

(The potential also goes to zero at infinity.) Given $\sigma(\mathbf{r})$, we can solve for

$\psi_j(\mathbf{r})$ and then calculate the current at the $i$-th conductor divided by $V_j$,

$$\begin{aligned} Y_{ij} &= -\frac{1}{V_j}\int \chi_i(\mathbf{r})\sigma(\mathbf{r})\vec{\mathbf{n}}\cdot\nabla\psi_{\mathbf{j}}(\mathbf{r})\mathbf{dS} \\ &= -\frac{1}{V_j}\int \sigma(\mathbf{r_i})\vec{\mathbf{n}}_{\mathbf{i}}\cdot\nabla\psi_{\mathbf{j}}(\mathbf{r_i})\mathbf{dS_i}; \end{aligned} \tag{16.3}$$

$\vec{n}$ is the unit normal; $Y_{ij}$ are the transfer admittances for the problem; subscript $i$ refers to integration over the $i$-th electrode.

Minerbo [5] used the following approach to compute the conductivity $\sigma(\mathbf{r})$. Set

$$\mathcal{L}_1(\sigma) = \sum_{i,j}|Y_{ij} - D_{ij}|^2/s_{ij}^2,$$

$$\mathcal{L}_2(\sigma) = \alpha^2\int|\nabla\log\sigma(\mathbf{r})|^2\mathbf{dr},$$

where $D_{ij}$ are the measured values of the transfer admittances $Y_{ij}$ and $s_{ij}$ are estimated standard deviations (that are basically known). Here $\alpha$ is adjustable positive parameter. He then seeks to

$$\text{find } \sigma(\mathbf{r}) \text{ which minimizes } \mathcal{L} = \mathcal{L}_1(\sigma) + \mathcal{L}_2(\sigma). \tag{16.4}$$

The reason for working with $\nabla(\log\sigma)$ rather than $\nabla\sigma$ is that the variation of the conductivity from the borehole mud to the rock formation is large; indeed, $\sigma_b/\sigma$ is up to 40000.

The minimization of $\mathcal{L}$ is done with a Marquardt-Levenberg method, which is a sort of Newton's method applied to the Euler equation (of the variational problem). Setting $q = \log(\sigma/\sigma_R)$ where $\sigma_R$ is a reference conductivity, the update of $q$ at the $k$-th iteration is obtained (after discretization) by solving the linear system of equations

$$\begin{gathered} \frac{1}{s_{ij}}\sum_{n,m}\sigma_{nm}^{k-1}\frac{\delta Y_{ij}(\sigma^{k-1})}{\delta\sigma_{nm}}(q_{nm}^k - q_{nm}^{k-1}) = \frac{1}{s_{ij}}\left[D_{ij} - Y_{ij}(\sigma^{k-1})\right], \\ \alpha\widehat{\nabla}^2 q_{nm}^k = 0, \end{gathered} \tag{16.5}$$

where $nm$ is the conductivity pixel, $\widehat{\nabla}$ is the finite-difference Laplacian, and $\delta Y_{ij}/\delta\sigma_{nm}$ is the sensitivity fuction (partial derivative) that describes the effect on $Y_{ij}$ of a change in the conductivity of the $nm$-th pixel.

To handle the large jump in conductivity between the borehole and formation the Laplacian $\widehat{\nabla}$ needs to be modified at the borehole wall. The way this is done in Minerbo [5] is to replace the discretized $\mathcal{L}_2$ by

$$\alpha^2\sum_m\sum_n \Delta r_n\Delta z_m\frac{r_n + r_{n-1}}{2}\left[\frac{(q_{nm} - q_{n-1,m})^2}{\Delta r_n^2} + \frac{(q_{nm} - q_{n,m-1})^2}{\Delta z_m^2}\right],$$

where, in $\sum_n$, the term $n$ corresponding to the borehole wall is dropped.

## 16.4 Sensitivity functions

The senstivity fuction $\delta Y_{ij}/\delta\sigma_{nm}$ can be obtained from the Frechet derivative,

$$\frac{\delta Y_{ij}}{\delta\sigma(\mathbf{r})} = \nabla\psi_i(\mathbf{r})\cdot\nabla\psi_{\mathbf{j}}(\mathbf{r}), \tag{16.6}$$

which gives the change in $Y_{ij}$ caused by a localized change (delta-function perturbation) in the conductivity at $\mathbf{r}$. This expression for the Frechet derivative is derived as follows [5]. The solution of the boundary value problem (16.1), (16.2) can be expressed in terms of a Green function that satisfies the equations,

$$\nabla\cdot(\sigma(\mathbf{r})\nabla\mathbf{G}(\mathbf{r},\mathbf{r}')) = \delta(\mathbf{r}-\mathbf{r}'), \quad \mathbf{r},\mathbf{r}' \text{ in } \boldsymbol{\Omega},$$

$$G(\mathbf{r},\mathbf{r}') = \mathbf{0}, \quad \mathbf{r} \text{ on } \mathbf{S}, \ \mathbf{r}' \text{ in } \boldsymbol{\Omega}.$$

Multiplying the equation for $G$ by $\psi$, the equation for $\psi$ (16.1) by $G$, subtracting, integrating over the domain $\Omega$, and applying the divergence theorem (and boundary conditions on $G$) give the representation

$$\begin{aligned}\psi_j(\mathbf{r}) &= V_j\int\chi_j(\mathbf{r}')\sigma(\mathbf{r}')\vec{\mathbf{n}}'\cdot\nabla\mathbf{G}(\mathbf{r}',\mathbf{r})\mathbf{dS}' \\ &= V_j\int\sigma(\mathbf{r_j})\vec{\mathbf{n}}_\mathbf{j}\cdot\nabla\mathbf{G}(\mathbf{r_j},\mathbf{r})\mathbf{dS_j}.\end{aligned}$$

The transfer admittance (16.3) is then given by

$$Y_{ij} = -\iint\sigma(\mathbf{r_i})(\vec{\mathbf{n}}_\mathbf{i}\cdot\nabla(\vec{\mathbf{n}}_\mathbf{j}\cdot\nabla\mathbf{G}(\mathbf{r_j},\mathbf{r_i})))\sigma(\mathbf{r_j})\mathbf{dS_j dS_i}. \tag{16.7}$$

To compute the Frechet derivative of $Y$, first compute the Frechet derivative of $G$ by adding a delta-function increment to $\sigma(\mathbf{r})$,

$$\nabla\cdot\sigma(\mathbf{r})\nabla\frac{\delta\mathbf{G}(\mathbf{r},\mathbf{r}')}{\delta\sigma(\mathbf{r}'')} = -\nabla\cdot(\delta(\mathbf{r}-\mathbf{r}'')\nabla\mathbf{G}(\mathbf{r},\mathbf{r}')), \quad \mathbf{r},\ \mathbf{r}',\ \mathbf{r}'' \text{ in } \Omega,$$

$$\frac{\delta G(\mathbf{r},\mathbf{r}')}{\delta\sigma(\mathbf{r}'')} = 0, \quad \mathbf{r} \text{ on } S, \ \mathbf{r}',\ \mathbf{r}'' \text{ in } \Omega.$$

The Green function $G$ can itself be used to solve this equation,

$$\begin{aligned}\frac{\delta G(\mathbf{r},\mathbf{r}')}{\delta\sigma(\mathbf{r}'')} &= -\int G(\mathbf{r},\mathbf{r_o})\nabla_\mathbf{o}\cdot(\delta(\mathbf{r_o}-\mathbf{r}'')\nabla_\mathbf{o}\mathbf{G}(\mathbf{r_o},\mathbf{r}'))\mathbf{dr_o} \\ &= \nabla''G(\mathbf{r},\mathbf{r}'')\cdot\nabla''\mathbf{G}(\mathbf{r}'',\mathbf{r}')\end{aligned} \tag{16.8}$$

This result and the equation (16.7) for the transfer admittance give the Frechet derivative (16.6).

## 16.5 Open problems

Although numerical simulations of the above scheme show that it can be effective in determining the conductivity of rock formations near the borehole, it does not model some important features of actual measurements and there are several areas where improvements are needed.

(i) The inversion can cope with very conductive mud (large $\sigma_b/\sigma$) provided the borehole radius is known in advance so that the smoothing constraint can be relaxed at the borehole wall.

*Question (1).* Is there a better way to handle the contrast $\sigma_b/\sigma$?

(ii) Contact impedance. Due to corrosion and other surface effects, each electrode has contact impedance $Z_j^{(c)}$ that varies slowly during the logging run. In fact, the contact impedance can vary both with time and with location along the electrode (as the electrode corrodes unevenly by exposure to the borehole mud). Contact impedance can be modeled as a thin surface layer (on the electrode) of thickness $d$ and impedance $Z_j^{(c)}/d$ as $d$ goes to zero [6]. There is a potential drop across this surface that depends on the amount of current $I_j$ passing through the electrode.

Thus, with contact impedance, if the potential supplied to the electrode by the driving electronics is $V_j$ and the potential that actually is applied to the medium just outside the electrode is $\psi_j$, then

$$\psi_j(\mathbf{r}) = \mathbf{f}(\mathbf{V_j}, \mathbf{c}(\mathbf{r}), \mathbf{I_j}),$$

where $c(\mathbf{r})$ is the boundary (contact) impedance and $f$ is a known function. In fact, a good model for contact impedance is the (Robin) boundary condition on $S_j$ [4]

$$\sigma(\mathbf{r})\vec{\mathbf{n}} \cdot \nabla\psi_{\mathbf{j}}(\mathbf{r}) = \frac{\mathbf{V_j} - \psi_{\mathbf{j}}(\mathbf{r})}{\mathbf{c}(\mathbf{r})}.$$

*Question (2).* Does the Dirichlet-to-Neumann map

$$V_j \longrightarrow I_j$$

determine both $\sigma(\mathbf{r})$ and $c(\mathbf{r})$ uniquely?

This is well known to be the case if there is no contact impedance; i.e., if $\psi_j \equiv V_j$ on the electrodes (and the electrodes cover the boundary of $\Omega$).

(iii) Resolution. How can one quantify the resolution (i.e., the smallest size change in conductivity that can be resolved) as a function of distance $r$ from the borehole axis?

(iv) The earth is 3D. How badly does the inversion do when inverting 3D data with 2D (axisymmetric) model?

(v) Optimal tool design. Given that the tool should map $\sigma$ out to a maximum distance $r_{\max}$ from the borehole, what is the optimal layout of electrodes?

One possible approach to Question (1) is the following:

Associate with each feasible borehole wall a representation: $\Gamma = \{r = r(z)\}$. Then solve (16.1), (16.2) with $\sigma(\mathbf{r})$ having jump discontinuity across the curve $\Gamma$, and compute the corresponding $Y_{ij}$. Now take the functional

$$\mathcal{L}(\sigma,\Gamma) = \mathcal{L}_1 + \alpha^2 \int_{\Omega\setminus\Gamma} |\nabla(\log\sigma(\mathbf{r}))|^{\mathbf{2}} \mathbf{dr} + \beta^{\mathbf{2}}|\mathbf{\Gamma}| \tag{16.9}$$

where $|\Gamma|$ is the length of $\Gamma$ and $\beta$ is a positive parameter, and choose $(\tilde{\sigma},\tilde{\Gamma})$ such that

$$(\tilde{\sigma},\tilde{\Gamma}) \text{ minimizes } \mathcal{L}(\sigma,\Gamma).$$

The functional (16.9) resembles the functional

$$J(f,\Gamma) = \mu^2 \int_{\Omega} |f-g|^2 + \int_{\Omega\setminus\Gamma} |\nabla f|^2 + \nu|\Gamma|$$

used by Mumford and Shah [7] for "cleaning up" images $g$ by introducing segmentation $\Gamma$ and smooth distribution function $f$ in sets $R_1,\ldots,R_k$ which are the components of $\Omega\setminus\Gamma$. Although it is not known whether the functional $J(f,\Gamma)$ has minimum, numerical examples give quite crisp images.

One could also introduce additional curves $\Gamma_1,\ldots\Gamma_N$ into the functional (16.9) to account for the jump that $\sigma(\mathbf{r})$ undergoes across transition layers.

## 16.6 References

[1] G. E. Archie, *The electrical resistivity log as an aid in determining some reservoir characteristics*, Trans. AIME, 146 (1942), 54–62.

[2] D.H. Davies, O. Faivre, M-T. Gounot, B. Seeman, J-C. Trouiller, D. Benimeli, A.E. Ferreira, D.J. Pittman, J-W. Smits, M. Randrianavony, B.I. Anderson, and J. Lovell, *Azimuthal Resistivity Imaging: A new generation laterolog*, SPE paper 24676, Proceedings 67th Annual Technical Conference and Exhibition of the Society of Petroleum Engineers (1992), 143–153.

[3] H.G. Doll, *The Laterolog: A new resistivity logging method with electrodes using an automatic focussing system*, Petroleum Transactions AIME, 192 (1951), 305–316.

[4] J.R. Lovell, *Finite element methods in resistivity logging*, Ph. D. thesis, Delft University of Technology (1993).

[5] G.N. Minerbo, *Resistivity mapping with an array laterolog*, Research Note EMG-002-94-08, Schlumberger-Doll Research (1994).

[6] K. Paulson, W. Breckon, and M. Pidcock, *Electrode modelling in electrical impedance tomography*, SIAM J. Appl. Math., 52 (1992), 1012–1022.

[7] D. Mumford and J. Shah, *Optimal approximations by piecewise smooth functions and associated variational problems*, Commun. Pure Appl. Math., 42 (1989), 576-685.

# 17

# A phenomenological model for case–II diffusion in polymers

Fluid absorbed in a polymer is called solvent. The automobile industry uses polymetric material in a number of applications. Polymeric-solvent interactions are important in plastic fuel tanks and fuel line components as well as in seals used to prevent leakage of lubricants or of water. Polymers can be in the glassy or in the rubbery state. Since the glassy state is more rigid at the molecular level, diffusion of solvent through glassy polymers is slower than diffusion through rubbery polymers. Diffusion of solvent in polymeric materials is accompanied by a variety of processes, including swelling, development and release of macroscopic elastic stresses, and changes of state. In particular, if, within an initially glassy polymeric material, the concentration of solvent exceeds some (material dependent) threshold concentration, the material becomes rubbery. This change of state is known as *plasticization*.

In order to improve the design of the polymeric components that come in contact with fuel, it is important to understand the phenomenon of plasticization and analyze its consequences. On May 20, 1996 Giuseppe Rossi from Ford Motor Company addressed the issue of providing a consistent description for the kinetics of the plasticization process and how it interferes with solvent transport, e.g., with diffusion. He described recent work [1] which develops a very simple model where the kinetics of plasticization controls the solvent flux at the interface between the glassy and the plasticized regions. This model qualitatively accounts for the available experimental evidence, including (i) the fact that both the distance covered by the interface and the solvent uptake are initially linear in time (case-II diffusion), and (ii) the existence of an induction period preceding the establishment of a moving interface.

The model in [1] is based on discretizing the space variable. Later on in this chapter we shall introduce the continuous version of the model, as more recently developed by A. Friedman and G. Rossi. The advantage of this continuous version is that it formulates very precisely the conditions on the free boundary separating the glassy and rubbery states. This formulation will be used to derive mathematical results on the shape of the free boundary.

## 17.1 Case–II diffusion

Figure 17.1 shows polymer in three states: glassy, rubbery (elastic), and melt (fluid). For dry polymer, the transition from glass to rubber occurs when the temperature $T$ is increased above the "glass transition temperature" $T_g$, and the transition from rubber to melt, e.g., the melting of crystals, occurs as the temperature is further increased above the "melting temperature" $T_m$.

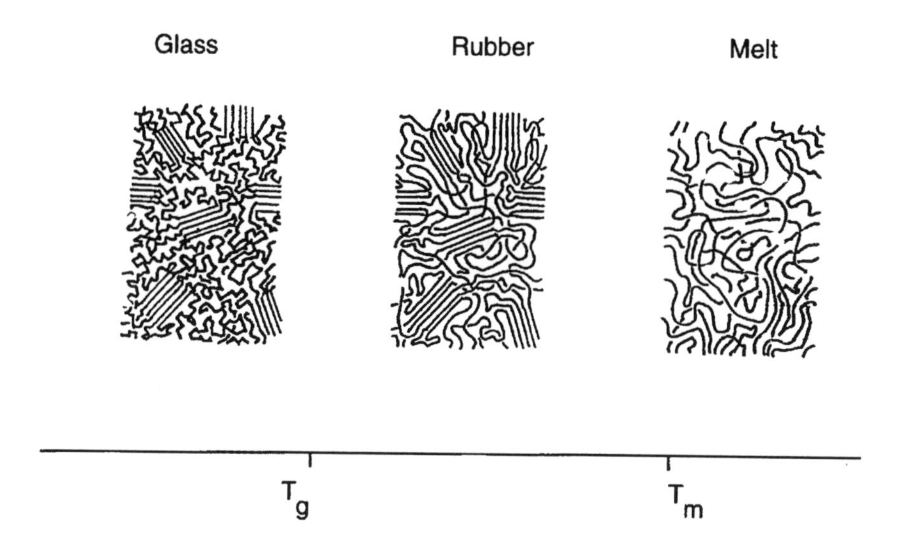


FIGURE 17.1.

Figure 17.2 shows in more detail the texture of glassy and rubbery states; here crystals are schematically shown as large dots. Since significant movement of chains with respect to each other is possible in the rubbery but not in the glassy state, solvent molecules can move more freely in the rubbery medium; in other words, the diffusion of solvent in glass is smaller than the diffusion of solvent in rubber.

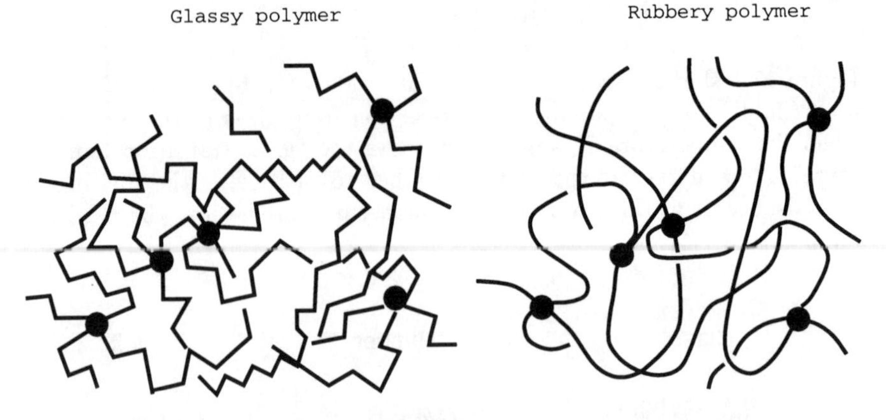


FIGURE 17.2.

Glassy polymer can change into rubbery state at fixed temperature by absorbing solvent. When the solvent concentration exceeds a threshold concentration $\widetilde{\varphi}$, the polymer becomes rubbery, i.e., it plasticizes.

The following has been observed [2]:

If we put a dry glassy polymer in fluid (Figure 17.3(a)) then the fluid diffuses according to Fick's law until time $\tau_0$ (the *induction* time). Thereafter a rubbery phase develops near the fluid boundary, and the sharp front between the rubbery and glassy states begins to advance (see Figure 17.3(b), (c)) at rate which is linear for times near $\tau_0$.

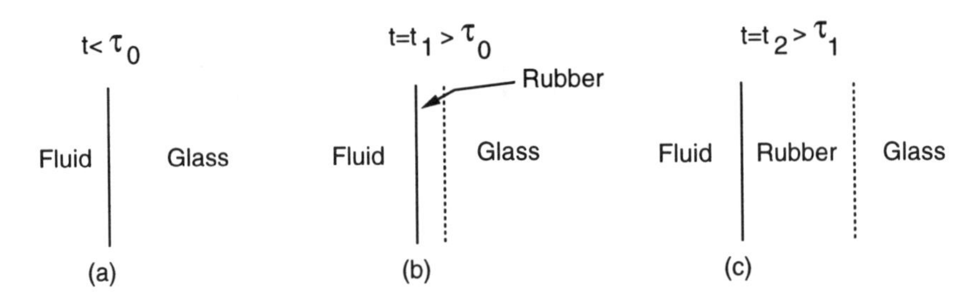


FIGURE 17.3.

The above behavior cannot be adequately explained by Fick's law, and one refers to this situation as case–II diffusion. Various models have been developed in order to explain the case–II diffusion; see [3, Chapter 4] [1] and the references therein. Another interesting aspect is the swelling that occurs in the rubbery state; this will not be discussed here, but the interested reader is referred to [3, Chapter 5], [4, Chapter 12] and the references therein. We also mention article [5] which includes a comprehensive bibliography.

## 17.2 The discretized model

Following [1] we make two assumptions:

(1) Plasticization occurs when the concentration $\varphi$ exceeds a threshold concentration $\widetilde{\varphi}$ ($\widetilde{\varphi}$ depends on the temperature $T$, but $T$ will be constant).

(2) The glass/rubber interface cannot advance with velocity larger than $v$, to be prescribed later on.

Consider the 1-d model and divide the x-axis into intervals of length $\Delta x$; denote the midpoint of the $i$-th interval by $x_i$, the concentration at $x_i$ by $\varphi_i$, and the flux from $x_i$ to $x_{i+1}$ by $f_i$; see Figure 17.4

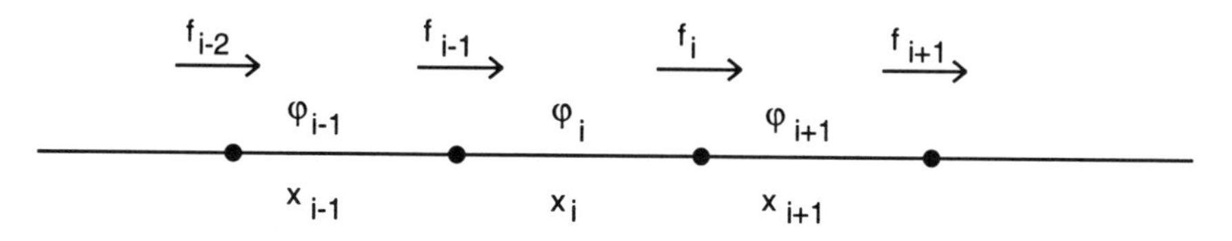


FIGURE 17.4.

If $\varphi_i > \varphi_{i+1} > \widetilde{\varphi}$ or $\varphi_{i+1} < \varphi_i < \widetilde{\varphi}$ then

$$f_i(t) = -D(\varphi_i)(\varphi_{i+1} - \varphi_i)/\Delta x \tag{17.1}$$

where $D(\varphi) = D_0$ in the glass and $D(\varphi) = D_1$ in the rubber; $D_0 < D_1$.

Consider next the case

$$\varphi_i \geq \widetilde{\varphi} \quad \text{and} \quad \varphi_{i+1} < \widetilde{\varphi} \ ,$$

i.e., the front is between $x_i$ and $x_{i+1}$. Suppose the interface moves with velocity $v$. Then $\varphi_{i+1}$ can exceed $\widetilde{\varphi}$ only after a time $\Delta x/v$. For smaller times $\delta t, \varphi_{i+1}$ will grow by an amount

$$\frac{\delta t}{\Delta x}(\widetilde{\varphi} - \varphi_{i+1})v \ .$$

Hence

$$f_i(t) = (\widetilde{\varphi} - \varphi_{i+1})v \ .$$

The following phenomenological law is assumed in [1]:

$$v = v(\varphi) = \frac{v_0\widetilde{\varphi}}{\widetilde{\varphi} - \varphi} \ , \quad \text{where } \varphi \text{ is the solvent concentration in glass.} \tag{17.2}$$

This implies that

$$f_i(t) = v_0\widetilde{\varphi} \ . \tag{17.3}$$

If the interface moves with velocity $\leq v$ then we similarly get

$$f_i(t) \leq v_0 \widetilde{\varphi} \ . \tag{17.4}$$

The reason for the assumption (2) above is that for the interface to advance it needs some time in order to "break down" the glassy structure near the interface. In deciding when to adopt (17.3) and when to adopt (17.4) we compare $v_0\widetilde{\varphi}$ with the flux given by (17.1), i.e., with

$$f_i(t) = -D_1(\varphi_{i+1} - \varphi_i)/\Delta x \ :$$

If

$$v_0\widetilde{\varphi} < -D_1(\varphi_{i+1} - \varphi_i)/\Delta x \quad \text{then we take} \quad f_i(t) = v_0\widetilde{\varphi} \ , \tag{17.5}$$

whereas if

$$v_0\widetilde{\varphi} \geq -D_1(\varphi_{i+1} - \varphi_i)/\Delta x \quad \text{then we take} \quad f_1(t) = -D_1(\varphi_{i+1} - \varphi_i)/\Delta x \ . \tag{17.6}$$

This means that as long as the flux on the rubbery side is larger than $v_0\widetilde{\varphi}$, the front moves with velocity $v$; once the flux on the rubbery side becomes smaller than $v_0\widetilde{\varphi}$, the front will move with velocity smaller than $v$, the restriction (2) is automatically satisfied, and the usual Fickian diffusion will take hold.

Next we impose initial condition

$$\varphi(x,0) = 0 \ , \ 0 < x < \infty \ , \tag{17.7}$$

i.e., the polymer is initially dry, and boundary conditions

$$-D_0\varphi_x(0,t) = v_0\widetilde{\varphi} \quad \text{if} \quad 0 < t < \tau_0 \ , \tag{17.8}$$

$$\varphi(0,t) = \Phi \quad \text{if} \quad \tau_0 < t < \infty \qquad (\Phi > \widetilde{\varphi}) \ . \tag{17.9}$$

We note that (17.8) is the condition that the front does not begin to advance until the concentration $\varphi(0,t)$ reaches the level $\widetilde{\varphi}$, i.e., $\tau_0$ is determined by

$$\varphi(0,\tau_0) = \widetilde{\varphi} \ . \tag{17.10}$$

We have three regimes. In regime 1, $0 < t < \tau_0$,

$$\varphi_t = D_0\varphi_{xx} \ , \ 0 < x < \infty \ , \ 0 < t < \tau_0$$

and we can easily solve for $\varphi$:

$$-\varphi_x(x,t) = \frac{\widetilde{\varphi}\, v_0}{D_0} \ \text{erfc} \ \left(\frac{x}{2\sqrt{D_0 t}}\right) \tag{17.11}$$

where

$$\text{erfc} \ z = \frac{2}{\sqrt{\pi}} \int_z^\infty e^{-\lambda^2} d\lambda \ ;$$

the induction time $\tau_0$ is given by

$$\tau_0 = \frac{\pi D_0}{4v_0^2} . \tag{17.12}$$

In regime 2, $\tau_0 < t < \tau_1$, (17.5) holds, and $\tau_1$ is the first time when

$$v_0\widetilde{\varphi} = -D_1(\varphi_{i+1} - \varphi_i)/\Delta x \quad \text{at the interface.} \tag{17.13}$$

Finally, in regime 3, $t > \tau_1$ and we use the standard Fick's law with continuity of concentration and flux across the interface. Thus case–II diffusion occurs just in the interval $\tau_0 < t < \tau_1$.

The numerical results obtained in [1] are qualitatively in agreement with experiments. Similar results concerning the transition to a diffusion limited regime were obtained by Astarita and Sarti [6]. However in [6] the authors impose a constitutive equation for the front velocity, instead of just the upper bound assumption imposed in the present model. Mathematical analysis of the free boundary problem formulated in [6] is given in [7].

## 17.3 The continuous model

A. Friedman and G. Rossi [8] have recently extended the discretized model of [1] to a continuous model. Here the main point is to formulate the case–II diffusion, where $\tau_0 < t < \tau_1$. Denote the front by

$$x = s(t) , \qquad \tau_0 < t < \tau_1 .$$

Then

$$\varphi_t = D_0\varphi_{xx} \quad \text{if} \quad s(t) < x < \infty , \tag{17.14}$$

$$\varphi(s(t)+, t) = \widetilde{\varphi} , \ \tau_0 < t < \tau_1 , \tag{17.15}$$

$$-D_0\varphi_x(s(t)+, t) = v_0\widetilde{\varphi} , \ \tau_0 < t < \tau_1 , \tag{17.16}$$

and

$$-\varphi_x(x, \tau_0) = \frac{\widetilde{\varphi}v_0}{D_0}\, erfc\frac{xv_0}{D_0\sqrt{\pi}} , \ s(\tau_0) < x < \infty . \tag{17.17}$$

This is a free boundary problem, which we shall consider more closely later on. Once a solution has been found, we can proceed to determine $\varphi(x, t)$ in the rubbery region by solving

$$\varphi_t = D_1\varphi_{xx} \quad \text{if} \quad 0 < x < s(t) , \tag{17.18}$$

$$\varphi(0, t) = \Phi , \ \tau_0 < t < \tau_1 , \tag{17.19}$$

$$s(\tau_0) = 0 , \tag{17.20}$$

and (the conservation of mass law)

$$-D_1\varphi_x(s(t)-,t)+D_0\varphi_x(s(t)+,t) = \dot{s}(t)(\varphi(s(t)-,t)-\widetilde{\varphi}) \text{ for } \tau_0 < t < \tau_1 . \tag{17.21}$$

Notice that when (17.13) first occurs, $-D_1\varphi_x(s(t)-,t)$ becomes equal to $v_0\widetilde{\varphi}$ and thus, by (17.16),

$$D_1\varphi_x(s(t)-,t) = D_0\varphi_x(s(t)+,t) , \tag{17.22}$$

or, by (17.21),

$$\varphi(s(t)-,t) = \widetilde{\varphi} . \tag{17.23}$$

Thus $\varphi(s(t)-,t) > \widetilde{\varphi}$ if $\tau_0 < t < \tau_1$, and $\varphi(s(t)-,t) = \widetilde{\varphi}$ at $t = \tau_1$. (At the end of this section we shall prove that $\tau_1 < \infty$.) The transition at time $t = \tau_1$ to regime 3 can consequently be defined as the first time that

$$\varphi(s(\tau_1)-,\tau_1) = \widetilde{\varphi} .$$

For $t > \tau_1$, the interface conditions are (17.22) and

$$\varphi(s(t)-,t) = \varphi(s(t)+,t) .$$

This means that for $t > \tau_1$ we solve the single parabolic equation

$$\varphi_t = (D(\varphi)\varphi_x)_x \qquad 0 < x < \infty , \quad t > \tau_1$$

(where $D(\varphi) = D_1$ if $\varphi > \widetilde{\varphi}$ , $D(\varphi) = D_0$ if $\varphi < \varphi_1$) and then the front is defined by

$$x = s(t) \quad \text{such that} \quad \varphi(s(t),t) = \widetilde{\varphi} .$$

We shall now concentrate on the free boundary problem (17.14)–(17.17). For simplicity we replace $t$ by $t + \tau_0$ and normalize the constants; this will not affect our results. The problem we consider is then to find a curve $x = s(t)$ and a function $\varphi(x,t)$ defined for $s(t) < x < \infty$ such that:

$$\varphi_t = \varphi_{xx} \quad \text{if} \quad s(t) < x < \infty , \ t > 0 , \tag{17.24}$$

$$\varphi(s(t),t) = B \quad \text{if} \quad t > 0 , \quad B > 0 , \tag{17.25}$$

$$-\varphi_x(s(t),t) = 1 \quad \text{if} \quad t > 0 , \tag{17.26}$$

$$-\varphi_x(x,0) = \frac{2}{\sqrt{\pi}}\int_x^\infty e^{-\lambda^2}d\lambda \quad \text{if} \quad 0 < x < \infty , \tag{17.27}$$

$$s(0) = 0 . \tag{17.28}$$

Since $\frac{2}{\sqrt{\pi}}\int_0^\infty e^{-\lambda^2}d\lambda = 1$, $\varphi_x(s(t),t) = \varphi_x(x,0)$ at $t = x = 0$. Set

$$u = 1 + \varphi_x .$$

By (17.26), $u = 0$ if $x = s(t)$. Differentiating (17.25) in $t$ and using (17.26) we also get $u_x = \dot{s}$ on $x = s(t)$. Hence the system (17.24)–(17.28) reduces to the following Stefan problem with negative latent heat:

$$
\begin{aligned}
&u_t = u_{xx} \quad \text{if} \quad s(t) < x < \infty \ , \ t > 0 \ , \\
&u = 0 \ , \ u_x = \dot{s} \quad \text{on the free boundary} \quad x = s(t) \ , \quad t > 0 \ , \qquad (17.29) \\
&u(x,0) = h(x) \quad \text{if} \quad 0 < x < \infty \ ,
\end{aligned}
$$

where

$$h(x) = 1 - \frac{2}{\sqrt{\pi}} \int_x^{\infty} e^{-\lambda^2} d\lambda = \operatorname{erf}(x) \ , \quad h(0) = 0 \ . \tag{17.30}$$

The function $U = -u$ satisfies the standard Stefan problem for supercooled water.

**Theorem 17.1** *The Stefan problem (17.29), (17.30) has a unique boundary solution with $s(t)$ continuously differentiable for some interval $0 \le t \le T$, and $\dot{s}(t) > 0$ for $0 < t \le T$.*

Indeed, this follows from general theory (see, for instance, [9]).
Clearly

$$s(t) \approx a + \dot{s}(0)t \quad \text{for small} \quad t \ ,$$

where $\dot{s}(0) = h'(a)$, and this gives the nearly linear velocity of the front in case–II diffusion, for small time.

**Theorem 17.2** $\dfrac{ds}{dt}$ *is strictly monotone increasing.*

*Proof.* Note that

$$h'(x) = \frac{2}{\sqrt{\pi}} e^{-x^2} > 0 \ , \ h''(x) = -\frac{4}{\sqrt{\pi}} x^2 e^{-x^2} < 0 \ ,$$

$$\frac{h''(x)}{h'(x)} = -2x \to -\infty \quad \text{if} \quad x \to \infty \ , \quad \text{and} \quad \left(\frac{h''}{h'}\right)' < 0 \ .$$

These conditions on the initial data enable us to extend the proof of Theorem 1.1 in [10] to the present case. That proof is based on analysis of the level lines of the function $v = u_t/u_x$, and it shows that $u_x > 0$ , $u_t < 0$ and level lines of $v$ initiating on the free boundary must end on the $x$-axis. Since, on the one hand, $v = -\dot{s}$ on the free boundary, and on the other hand $v = h''/h'$ on the $x$-axis, $\dot{s}(t)$ is strictly monotone increasing.

Using Theorem 17.2 we prove:

**Theorem 17.3** *The Stefan problem (17.29), (17.30) has a unique solution for all $t > 0$, and $\dot{s}(t)$ in positive and strictly monotone increasing for all $t > 0$.*

*Proof.* If the assertion is not true then there is a finite time $T$ such that either

$$\text{(i)}\ s(T) \equiv s(T-) < \infty \quad \text{but} \quad \dot{s}(T) \equiv \dot{s}(T-0) < \infty\ , \quad \text{or}$$

$$\text{(ii)}\ s(T) = \infty\ ;$$

indeed, otherwise, $s(T) < \infty$ and $\dot{s}(T) < \infty$, and the proof of Theorem 17.1 can be applied to continue the solution beyond $t = T$.

We shall now rule out both cases (i) and (ii).

Suppose (i) holds and introduce the rectangles

$$R\ = \{\ (x,t)\ ;\quad s(T)+2 < x < s(T)+3\ ,\quad 0 < t < T\}\ .$$

$$R_1 = \{\ (x,t)\ ;\quad s(T)+1 < x < s(T)+4\ ,\quad 0 < t < T\}\ .$$

The function $u$ satisfies $0 < u < 1$, $u_x > 0$, $u_t < 0$ in the larger rectangle $R_1$, and therefore $u$ can be continued into $\overline{R}$, as a solution of the heat equation and, furthermore, $0 < u < 1$, $u_x > 0$, $u_t < 0$ in $\overline{R}$ (by the strong maximum principle). In particular, the function $v = u_t/u_x$ is bounded in $\overline{R}$. However, on the level lines of $v$ starting near $(s(T),T)$ $|v|$ takes arbitrarily large values (since $|v| = \dot{s}(t) \to \infty$ if $t \to T$). Hence these level lines cannot cross $R$, and must therefore intersect the $x$-axis at points $(x,0)$ with $x < s(T)+2$; this is a contradiction since $v = h''(x)/h'(x)$ is uniformly bounded for such $x$.

Suppose next that case (ii) holds. We introduce the translates

$$u_\lambda(x,t) = u(x+\lambda,\ t)\ , \quad \lambda > 0\ .$$

As $\lambda \to +\infty$ the $u_\lambda$ converge to a solution $w$ of the heat equation in the strip

$$S = \{-\infty < x < \infty,\quad 0 < t < T\}\ ,\quad \text{and} \quad 0 \le w \le 1 \quad \text{in } S\ ,$$

$$w(x,0) \equiv 1 \quad (\text{since} \quad h(x) \to 1 \quad \text{if} \quad x \to \infty)\ ,$$

$$w(x,1) \equiv 0 \quad (\text{since} \quad u(s(t),t) = 0\ ,\ s(t) \to \infty \quad \text{if} \quad t \to T)\ .$$

By backward uniqueness for non-negative solutions of the heat equation (see, for instance, [11]), we must have $w \equiv 0$, a contradiction.

We finally prove:

**Theorem 17.4** $\tau_1 < \infty$ .

*Proof.* If this is not true then

$$\varphi(s(t)- \, , t) > \widetilde{\varphi} \quad \text{for all} \quad t > \tau_0 \tag{17.31}$$

and, by (17.21), (17.16),

$$D_1 \varphi_x(s(t)- \, , t) \leq D_0 \varphi_x(s(t)+ \, , t) = -v_0 \widetilde{\varphi} \quad \text{if} \quad t > \tau_0 \ .$$

We compare $\varphi$ in the domain

$$\Omega = \{0 < x < s(t) \ , \quad t > \tau_0\}$$

with the solution $\psi$ of

$$\psi_t = D_1 \psi_{xx} \quad \text{in} \quad \Omega \ ,$$

$$\psi(0,t) = M \ , \quad t > \tau_0 \ ,$$

$$\psi_x(x, \tau_0) = k(x) \ , \quad 0 < x < s(\tau_0) \ ,$$

$$D_1 \psi_x(s(t)- \, , t) = -v_0 \widetilde{\psi}$$

where $M$ and $k(x)$ are chosen such that

$$k(0) = M > \Phi \ , \quad k(x) \geq \psi(x, \tau_0) \ , \quad D_1 k'(x) \leq -v_0 \widetilde{\varphi} \ .$$

By the maximum principle, $\varphi \leq \psi$ in $\Omega$. On the other hand the maximum principle can be used to deduce that $D_1 \psi_x \leq -v_0 \varphi$ in $\Omega$ and, consequently,

$$\varphi(s(t)- \, , t) \leq \psi(s(t), t) \to -\infty \quad \text{if} \quad t \to \infty \ ,$$

which is a contradiction.

## 17.4 References


[1] G. Rossi, P.A. Pincus and P.-G. De Gennes, *A phenomenological description of case-II diffusion in polymeric materials*, Europhysics Letters, 32 (1995), 391–396.

[2] T.P. Gall, R.C. Lasky and E.J. Kramer, *Case II diffusion: effect of solvent molecule size*, Polymer, 31 (1990), 1491–1499.

[3] A. Friedman, *Mathematics in Industrial Problems, Part 5*, IMA Volume 49, Springer–Verlag, New York (1992).

[4] A. Friedman, *Mathematics in Industrial Problems, Part 3*, IMA Volume 31, Springer-Verlag, New York (1990).

[5] M.A. Samus and G. Rossi, *Methanol absorption in ethylene-vinyl alcohol copolymers: Relations between solvent diffusion and changes in glass transition temperature in glassy polymeric materials*, Macromolecules, 29 (1996), 2275–2288.

[6] G. Astarita and G.C. Sarti, *A class of mathematical models for sorption of swelling solvents in glassy polymers*, Polym. Eng. Sci., 18 (1978), 388–395.

[7] A. Fasano, G.H. Meyer and M. Primicerio, *On a problem in the polymer industry: Theoretical and numerical investigation of swelling*, SIAM J. Math. Anal., 17 (1986), 945–960.

[8] A. Friedman and G. Rossi, *Phenomenological continuum equations to describe case II diffusion in polymeric materials*, Macromolecules, to appear.

[9] A. Friedman, *Partial Differential Equations of Parabolic Type*, Prentice–Hall, Englewood Cliffs, N.J. (1964).

[10] A. Friedman and R. Jensen, *Convexity of the free boundary in the Stefan problem and in the dam problem*, Archieve Rat. Mech. Anal., 67 (1977), 1–24.

[11] D.V. Widder, *The Heat Equation*, Academic Press, New York (1975).

# 18

# Nonlinear elastic-viscoelastic correspondence

The term "elastomer" is applicable to all amorphous (rubber-like) polymeric solids (natural or synthetic) that can sustain large deformations with relatively small applied force and which recover their original shape within a relatively short time after the force is removed. The modern automobile has more than 600 elastomeric components, many of which are critical load-bearing structural components. There is a need for accurate mechanical analysis of both the design and performance of such components. This requires solving nonlinear viscoelastic boundary value problems. On June 21, 1996 Kenneth N. Morman, Jr. from Ford Motor Company presented correspondence principles for a class of compressible, isotropic and homogeneous viscoelastic rubber-like materials. These principles provide relatively easy and computationally efficient method for obtaining the nonlinear viscoelastic solution from the nonlinear elastic solution, instead of directly solving the viscoelastic problem. He illustrated the utility of the correspondence principles with several examples. The principles are established under certain assumptions on the deformation; extending the validity of these principles to more general deformations remains a challenging problem.

## 18.1 Automotive elastomeric components

Automotive elastomeric components include:

(i) Suspension bushings[1];

(ii) Body and engine mounts;

(iii) Grommets or rubber washers;

(iv) Exhaust system hangers;

(v) Seals and gaskets;

(vi) Hoses.

---

[1]A removable cylindrical part for an opening (as of a mechanical part) used to minimize friction and serve as a guide for the relative movement between two mechanical parts.

Many components are used for load bearing which are responsive for support, safety, handling and comfort. Effective performance requires qualities such as noise attenuation, strength and durability, shock attenuation, vibration isolation, and sealing.

Figure 18.1 shows various suspension bushings and shock isolators used in an automobile.

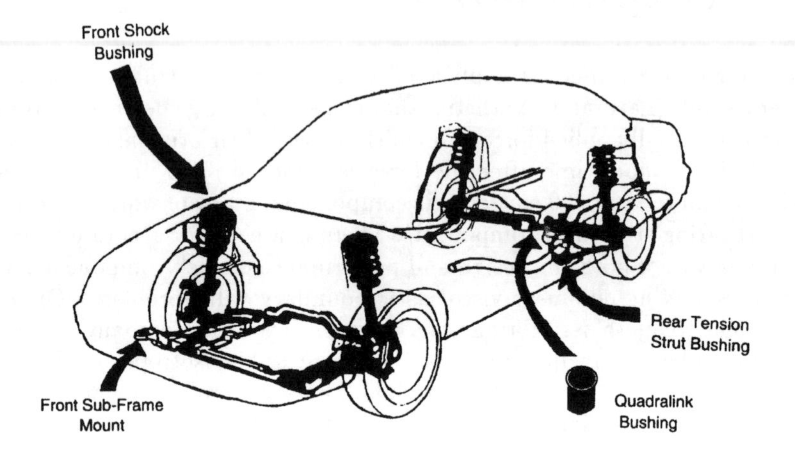


FIGURE 18.1.

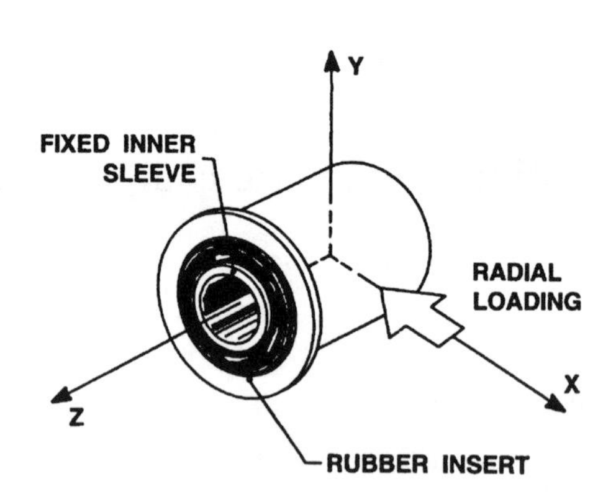


Quadralink bushing under radial loading

FIGURE 18.2.

The quadralink bushing, which consists of a metal inner sleeve, a metal outer-sleeve and rubber insert, is depicted in Figure 18.2. The bushing is assembled by forcing the rubber insert between the two sleeves with the aid of special lubricants. In a typical service environment, this bushing may be subjected to combined axial, radial, conical, and torsional loadings. The

The quadralink bushing, which consists of a metal inner sleeve, a metal outer-sleeve and rubber insert, is depicted in Figure 18.2. The bushing is assembled by forcing the rubber insert between the two sleeves with the aid of special lubricants. In a typical service environment, this bushing may be subjected to combined axial, radial, conical, and torsional loadings. The non-linear viscoelastic response behavior of this bushing to two different radial deflection histories is depicted in Figures 18.3 and 18.4.

In Figurc 18.3, wc scc that cach stcp in dcflcction input is accompanied by a relaxation in force with time to maintain that deflection step.

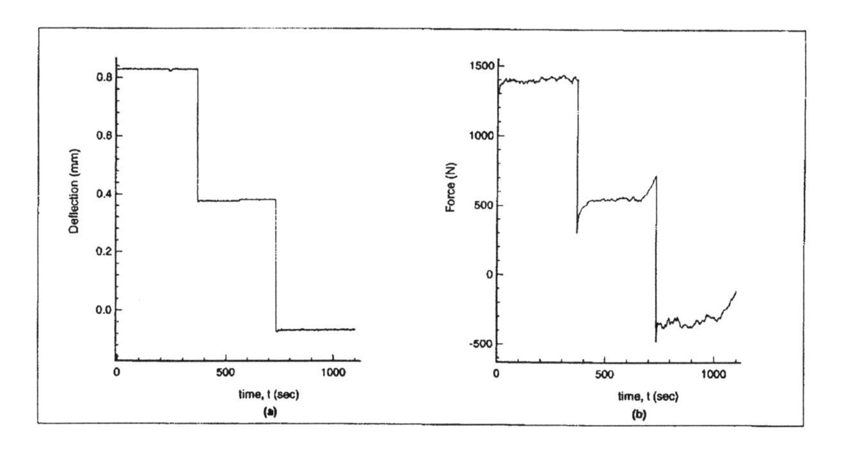


Radial Multiple-step deflection input and force response of the quadralink bushing

FIGURE 18.3.

Figure 18.4 shows that, for harmonic displacement input, the resulting force vs displacement is not a simple curve (as it would be expected for a purely elastic material), but a hysteresis loop.

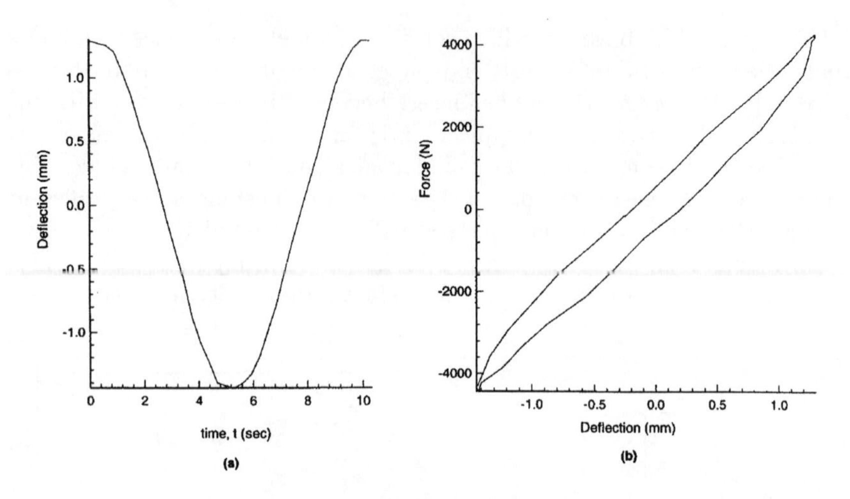


Radial Sinusoidal deflection wave input and force response of the quadralink bushing

FIGURE 18.4.

In the following sections we shall consider boundary value problems (BVP) for determining the nonlinear viscoelastic response of elastomeric components.

## 18.2 The constitutive equation

Deformation of continuum is described by

$$\mathbf{x} = \chi(\mathbf{X}, t)$$

where $\mathbf{X}$ is the position of particle at time $t = 0$ and $\mathbf{x}$ is its position at time $t$. We first consider nonlinear elastic material. The deformation gradient is defined as

$$\mathbf{F}(\mathbf{X}, t) = \text{Grad}_{\mathbf{X}}\, \chi(\mathbf{X}, t)\,, \quad \det \mathbf{F} > 0\,. \tag{18.1}$$

The left Cauchy–Green deformation measure is defined as

$$\mathbf{B} = \mathbf{F} \cdot \mathbf{F}^T. \tag{18.2}$$

The invariants of the deformation measure (under rigid motion) are

$$I_1(\mathbf{B}) = tr\mathbf{B}\ ,$$

$$I_2(\mathbf{B}) = \frac{1}{2}(I_1^2 - tr\mathbf{B}^2)\ ,$$

$$I_3 = \det \mathbf{B} = (\det \mathbf{F})^2\ .$$

For hyperelastic behavior, the stress tensor $\mathbf{T}$ satisfies the constitutive equation

$$\mathbf{T} = \mathbf{T}_e(\mathbf{B}) = h_0\mathbf{I} + h_1\mathbf{B} + h_2\mathbf{B}^2 \tag{18.3}$$

where

$$h_0 = \frac{2I_3}{\det \mathbf{F}}\frac{\partial W}{\partial I_3}\ , \quad h_1 = \frac{2}{\det \mathbf{F}}\frac{\partial W}{\partial I_1} + I_1\frac{\partial W}{\partial I_2}\ , \quad h_2 = -\frac{2}{\det \mathbf{F}}\frac{\partial W}{\partial I_2} \tag{18.4}$$

and $W$ is the strain energy density functional, usually determined from experiments. We refer to [1] for the general theory of nonlinear elasticity.

Consider next viscoelastic material. In this case we need to introduce also the right Cauchy–Green deformation measure

$$\mathbf{C} = \mathbf{F}^T \cdot \mathbf{F}\ .$$

Since the material has memory, we also introduce the *history* deformation gradient

$$\mathbf{F}^t(s) = \mathbf{F}(\mathbf{X}, t-s) \quad \text{for} \quad 0 \le s < \infty$$

(it is *current* if $s = 0$), and the *relative* deformation gradient

$$\mathbf{F}_t(\tau) = \mathbf{F}(\tau)\cdot\mathbf{F}(t)^{-1}\ , \qquad 0 \le \tau \le t\ .$$

The relative right Cauchy–Green deformation measure is defined by

$$\mathbf{C}_t(\tau) = \mathbf{F}_t^T(\tau)\cdot\mathbf{F}_t(\tau)\ ;$$

note that $\mathbf{C}_t(0) = \mathbf{B}^{-1}\ ,\ \mathbf{C}_t(\tau)|_{\tau=t} = \mathbf{I}$ .

The invariants of $\mathbf{C}$ are clearly the same as the invariants of $\mathbf{B}$.

Let $K(t)$ denote the bulk relaxation and $G(t)$ the shear relaxation under simple uniaxial tension. Motivated by work of Coleman and Noll [2], Morman [3] developed the following constitutive equation for rubber-like viscoelasticity:

$$\begin{aligned}\mathbf{T}(t) = \mathbf{T}_0(\mathbf{B}) &+ \frac{1}{3}\,\mathbf{I}\int_0^\infty \left[\frac{K'(s)}{K_e} - \frac{G'(s)}{G_e}\right] tr\mathbf{T}_t^e(t-s)ds \\ &+ \int_0^\infty \frac{G'(s)}{G_e}\ \mathbf{T}_t^e(t-s)ds\ .\end{aligned} \tag{18.5}$$

Here $\mathbf{T}_0(\mathbf{B})$ is the short-time response obtained by scaling long-time response, and is given by

$$\mathbf{T}_0(\mathbf{B}) = \frac{1}{3}\,\mathbf{I}\left[\frac{K_0}{K_e} - \frac{G_0}{G_e}\right] tr\mathbf{T}_e(\mathbf{B}) + \frac{G_0}{G_e}\ \mathbf{T}_e(\mathbf{B}) \tag{18.6}$$

where $\mathbf{T}_e(\mathbf{B})$ is defined in (18.3) ($\mathbf{T}_e(\mathbf{B})$ represents *long-time* response governed by hyperelastic behavior). The $\mathbf{T}_t^e$ in (18.5) is defined by

$$\mathbf{T}_t^e(\tau) = \mathbf{T}_t^e(\mathbf{M}_t(\tau)) = h_0(\tau)\mathbf{I} + h_1(\tau)\mathbf{M}_t(\tau) + h_2(\tau)(\mathbf{M}_t(\tau))^2 \tag{18.7}$$

where

$$\mathbf{M}_t(\tau) = \mathbf{B}^{1/2} \cdot \mathbf{C}_t(\tau) \cdot \mathbf{B}^{1/2}$$

and the $h_i(\tau)$ are the $h_i$ in (18.4) at time $\tau$. In the above,

$$K_0 = K(0),\ G_0 = G(0),\ K_e = K(\infty), \quad \text{and} \quad G_e = G(\infty)\ .$$

## 18.3 The nonlinear boundary value problem

Let $\Omega$ be a 3-d domain. The nonlinear elastic boundary value problem (NLEBVP) consists of the following of equations:

Conservation of mass

$$\det \mathbf{F} = \frac{\rho_0}{\rho} \quad \text{in} \quad \Omega \qquad (\rho = \text{density}); \tag{18.8}$$

Balance of momentum

$$\operatorname{div} \mathbf{T} + \rho \mathbf{b} = \rho \frac{D\mathbf{v}}{Dt}\ , \quad \mathbf{T} = \mathbf{T}^T \quad \text{in} \quad \Omega \tag{18.9}$$

where $\mathbf{b}$ is the body force, $\mathbf{v}$ the velocity, superscript $T$ denotes the transpose of a matrix, and $D/Dt$ denotes the material derivative;

Constitutive equation (assuming hyperelastic behavior)

$$\mathbf{T} = h_0 \mathbf{I} + h_1 \mathbf{B} + h_2 \mathbf{B}^2 \quad \text{in} \quad \Omega\ , \tag{18.10}$$

and the boundary conditions

$$\begin{aligned} \mathbf{t} = \mathbf{T} \cdot \mathbf{n} = \mathbf{t}_b(\mathbf{X}, t) \quad &\text{on } \partial\Omega_\sigma \text{ (prescribed traction)}, \\ \mathbf{x} = \chi(\mathbf{X}, t) = \chi_b(\mathbf{X}, t) \quad &\text{on } \partial\Omega_u \text{ (prescribed shape)} \end{aligned} \tag{18.11}$$

where $\partial\Omega = \partial\Omega_\sigma \cup \partial\Omega_u$ (disjoint union), and $\mathbf{t}$ is the traction.

For the nonlinear viscoelastic boundary value problem (NLVEBVP) (18.8), (18.9) and (18.11) are the same, but the constitutive law is the one given by (18.5).

## 18.4 Correspondence principles

Correspondence principles are powerful tools for solving viscoelastic boundary value problems. They express the solution in terms of the corresponding solution of the elastic boundary value problem. For correspondence principles in linear viscoelasticity we refer to [4] and the references therein. Correspondence principles for nonlinear viscoelasticity, in problems of fracture analysis, were given by Schapery [5].

In [6] Morman established correspondence principles between incompressible elastic and viscoelastic material solids. The present section describes his extension of these principles to the compressible case. We make the following three assumptions:

($A_1$) Shear relaxation modulus and bulk relaxation modulus are equal

$$\frac{G(t-s)}{G_e} \equiv \frac{K(t-s)}{K_e} \ ;$$

this implies that in (18.5)

$$\mathbf{T}(t) = \varphi(0)\mathbf{T}_e(\mathbf{B}) + \int_0^\infty \varphi'(s)\mathbf{T}_t^e(t-s)ds \ , \ \varphi(s) = \frac{G(s)}{G_e} \ . \qquad (18.12)$$

($A_2$) The motion of each material point at $\mathbf{X}$ of a non-homogeneously deformed solid up to time $t$ is a *sheared extension* with orthonormal basis $\mathbf{h}_i$ independent of $t-s$.

This concept, introduced by Colemen [7], means that

$$\mathbf{F}(t-s) = P(t-s)\mathbf{N}(t-s) \qquad (0 \leq s < \infty)$$

where $P(t-s)$ is orthogonal matrix for each $s$ and

$$\mathbf{N}(t-s) = \begin{pmatrix} \beta_1(t-s) & 0 & 0 \\ \zeta(t-s) & \beta_2(t-s) & 0 \\ 0 & 0 & \beta_3(t-s) \end{pmatrix} \ , \ \beta_i(t-s) > 0$$

with respect to the *canonical basis* $\mathbf{h}_i$. Another way of stating it is that the principal eigenvectors $\mathbf{u}_i(t-s)$ of $\mathbf{C}(t-s)$ satisfy:

$$\mathbf{u}_1(t-s) = \cos\alpha_d^t(s)\mathbf{u}_1(t) + \sin\alpha_d^t(s)\mathbf{u}_2(t) \ ,$$

$$\mathbf{u}_2(t-s) = -\sin\alpha_d^t(s)\mathbf{u}_1(t) + \cos\alpha_d^t(s)\mathbf{u}_2(t) \ ,$$

$$\mathbf{u}_3(t-s) = \mathbf{h}_3$$

where $\mathbf{u}_1(t), \mathbf{u}_2(t), \mathbf{h}_3$ are the eigenvectors of $\mathbf{C}(t)$ at time $t$ and

$$\alpha_d^t(s) = \alpha(t-s) - \alpha(t) \ , \qquad 0 \leq s < \infty \ .$$

Here we denote by $\alpha(t)$ the angle between the $\mathbf{h}_1$-axis and the principal direction of $\mathbf{C}(t)$ corresponding to the principle eigenvalue, measured counterclockwise. This motion occurs, for example, if $\mathbf{P}(t-s) = \mathbf{I}$, and

$$x^1 = x^1(X^1, t), x^2 = x^2(X^1, X^2, t), \ x^3 = x^3(X^3, t), \qquad -\infty < t < \infty \ .$$

($A_3$) Zero body and inertia forces.

This implies that

$$\text{div } \mathbf{T} = 0$$

so that (18.12) gives

$$\text{div } \mathbf{T}_e(\mathbf{B}) = -\text{div} \int_0^\infty \frac{\varphi'(s)}{\varphi(0)} \, \mathbf{T}_t^e(t-s)ds \; . \tag{18.13}$$

In general equation (18.13) represents a system of three simultaneous integro-partial differential equations that must be solved to determine the position $\mathbf{X}(t)$. For a certain broad class of motions, namely sheared extensions, equation (18.13) is satisfied identically since

$$\text{div } \mathbf{T}_e(\mathbf{B}) = -\text{div} \int_0^\infty \frac{\varphi'(s)}{\varphi(0)} \, \mathbf{T}_t^e(t-s)ds = 0 \; ;$$

the proof is too lengthy for presentation here. This result leads to three very useful correspondence principles which will be stated without proof in the following.

*Corresponding Principle I.* Let arbitrary shape $\chi_b(\mathbf{X}, t)$ on the boundary $\partial\Omega_u(t)$, with zero traction elsewhere, be specified such that $\partial\Omega_u(t') \subseteq \partial\Omega_u(t)$ if $t' \leq t$. Then the viscoelastic solution which satisfies the NLVE-BVP, under the assumptions $(A_1)$–$(A_3)$, has stress tensor

$$\begin{aligned} \mathbf{T}(t) &= \varphi(0)\mathbf{T}^e(\mathbf{B}) + \int_0^t \varphi'(s)\mathbf{T}_t^e(t-s)ds \; , \\ \mathbf{B} &= \mathbf{B}(\mathbf{X}, t) = \mathbf{B}^e(\mathbf{X}, t) \quad \text{in} \quad \Omega \; , \end{aligned} \tag{18.14}$$

where $\mathbf{B}^e, \mathbf{T}^e(\mathbf{B})$ and $\mathbf{T}_t^e(t-s) \quad (0 \leq s < \infty)$ are determined from the solution of the NLEBVP with

$$\chi^e(\mathbf{X}, t) = \chi_b(\mathbf{X}, t) \quad \text{on} \quad \partial\Omega_u(t)$$

and zero traction elsewhere.

The shear creep compliance $J(\tau)$ and the bulk creep compliance $B(\tau)$ are defined from the creep formulation of the stress/strain relation. What we shall need here is just the relation between these moduli and between the relaxation moduli:

$$\int_{-\infty}^t G(t-s)\dot{J}(\tau)d\tau = H(t) \; , \quad \int_{-\infty}^t K(t-\tau)\dot{B}(\tau)d\tau = H(t)$$

where $H(t)$ is the Heaviside function.

*Correspondence Principle II.* Let arbitrary traction $\mathbf{t}_b(\mathbf{X},t)$ on the boundary $\partial\Omega_\sigma(t)$ be specified such that $\partial\Omega_\sigma(t') \subseteq \partial\Omega_\sigma(t)$ if $t' \leq t$, with $\chi_b(\mathbf{X},t) = \mathbf{X}$ elsewhere on $\partial\Omega$. The viscoelastic solution of the NLVE-BVP, under the assumptions $(A_1) - (A_3)$, has stress tensor (18.13) where $\mathbf{B}^e, \mathbf{T}^e(\mathbf{B})$ and $\mathbf{T}_t^e(t-s)$ are determined from the solution of the NLEBVP with the traction

$$\mathbf{t}_b^e(\mathbf{X},t) = \psi(0)\mathbf{t}_b(\mathbf{X},t) + \int_0^\infty \psi'(s)\mathbf{F}_t(t-s)^T \cdot \mathbf{t}_b(\mathbf{X},t-s)\frac{da(t-s)}{da}\, ds \quad \text{on} \quad \partial\Omega_u(t)\,, \tag{18.15}$$

$$\psi(s) = \frac{J(s)}{J_e}\,.$$

Here $da(t-s)$, $da$ denote elements of surface area on $\partial\Omega$ at times $t-s$, and $t$, respectively.

*Correspondence Principle III.* Here $\chi_b(\mathbf{X},t)$ is prescribed on $\partial\Omega_u(t)$, $\mathbf{t}_b(\mathbf{X},t)$ is prescribed on $\partial\Omega_\sigma(t)$, where $\partial\Omega_u(t), \partial\Omega_\sigma(t)$ are independent of $t$. The stress tensor for the NLVEBVP, under the assumptions $(A_1)$–$(A_3)$, is given again by (18.14) with $\mathbf{T}^e(\mathbf{B})$ determined by solving the NLEBVP with the same $\chi_b(\mathbf{X},t)$ on $\partial\Omega_u$ and with $\mathbf{t}_b^e(\mathbf{X},t)$ as in (18.15).

Finite element analysis in the design of automotive elastomeric components has been carried out in [8]. The correspondence principles allow faster way of computing the stress. Morman used it to compute the strain in traction of a thick walled rubber tube, and he obtained the same results as are obtained from a well known closed form (analytic) solution. He also computed the stiffness in the shear/compression engine mount (Figure 18.5) and obtained results which are in complete agreement with the finite-element results in [8].

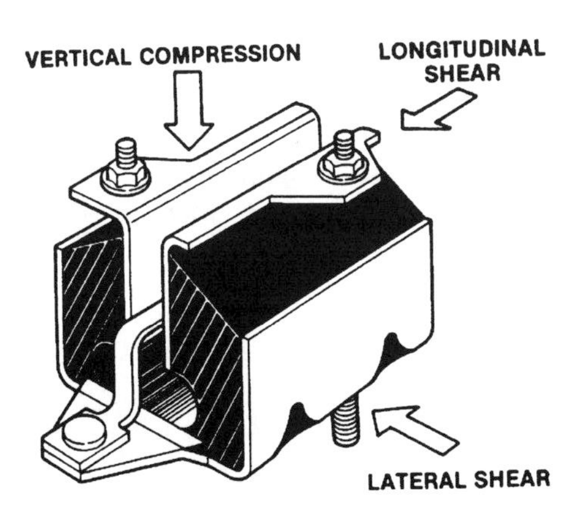


FIGURE 18.5.

The correspondence principles permit the determination of viscoelastic stress at preselected points within the body without having to solve the complete nonlinear viscoelasticity BVP. However so far these principles have been obtained only under the assumptions $(A_1)$–$(A_3)$.

The assumption $(A_1)$ is satisfied for many rubber-like materials, and $(A_3)$ is satisfied for a large class of deformations. Therefore the main question is:

Can the correspondence principles be extended to deformations which are not sheared extension motion?

## 18.5 References


[1] P.G. Ciarlet, *Mathematical Elasticity*, Vol. 1: Three Dimensional Elasticity,North-Holland, Amsterdam (1993).

[2] B.D. Coleman and W. Noll, *Foundations of linear viscoelasticity*, Rev. Mod Phys., 33 (1961), 329–249.

[3] K.N. Morman, *An adaptation of finite linear viscoelasticity theory for rubber-like viscoelasticity by use of a generalized strain measure*, Rheologica Acta, 27 (1988), 3–14.

[4] G.A. Graham, *The generalized partial correspondence principle in linear viscoelasticity*, Quart. Appl. Math., 46 (1988), 527–538.

[5] R.A. Schapery, *Correspondence principles and a generalized J integral for large deformation and fracture analysis of viscoelastic media*, Inter. J. of Fracture, 25 (1984), 195–223.

[6] K.N. Morman, *Non-linear elastic-viscoelastic correspondence principles for the solution of boundary-value problems involving a class of viscoelastic rubber-like solids*, Paper No.5 presented at a meeting of the Rubber Division, American Chemical Society, Montreal, Quebec, May 26-29, 1987: abstract, Rubber Chemistry and Technology, **60** (1987), p. 782.

[7] B.D. Coleman, *On the use of symmetry to simplify the constitutive equations of isotropic materials with memory*, Proc. Roy. Soc. A. 306 (1968), 449–476.

[8] K.N. Morman, *Application of finite-element analysis in the design of automotive elastomeric components*, Rubber Chemistry and Technology, 61 (1988), 503–533.

# 19

# Solutions to problems from previous parts

## 19.1 Part 1

The coating flow problem posed in Part 1, Chapter 3 has been studied by A. Friedman and J.J.L. Velázquez. They assume the no-slip condition up to the point of contact. This implies that the free boundary starts tangentially to the substrate. For the stationary problem they established existence and uniqueness [1] provided the flow is near a uniform flow. Planning to extend these results to non-stationary flows, they studied in [2] the linearized problem for the non-stationary problem and established existence and a priori estimates that will be used in the full nonlinear problem. The results in [2] require new Liouville-type theorems for fourth-order elliptic equations in a half-space [3].

## 19.2 Part 4

Chapter 4 dealt with limited coalescence problem for emulsion mixed with surfactant material. Mathematical results on existence, uniqueness, asymptotic behavior and a possible computational approach were obtained by O. Bruno, A. Friedman and F. Reitich [4]. The mathematical model is due to D. Ross and T.H. Whitesides who recently [5] reported experimental results as well as numerical results based on Monte Carlo simulations. Their work suggests interesting questions of mathematical content.

## 19.3 Part 7

Chapter 6 discussed the head-media interaction in magnetic tape. This is modelled as a system of two nonlinear elliptic equations for the tape (fourth order equation) and for the pressure (Reynolds' equation) in the space between the tape and head; the solution must be such that the spacing $h$ is $> 0$. Experimental and numerical results (in the 1-d case) indicate a boundary layer behavior for $h$ where the tape goes out of being above the head. A. Friedman and B. Hu [6] have recently analyzed the PDE system and established rigorously, for the 1-d case, the existence of a boundary layer.

Their method can be used (with no rigorous proof, though) to compute the boundary layers also for the 2-d case.

Chapter 10 discussed an approach by B. Morton to computational algorithm of mechanisms. A recent paper by B. Morton and M. Elgersma [7] develops computational techniques for analysis of 1-DOF, 7R spatial mechanisms.

## 19.4 References


[1] A. Friedman and J.J.L. Velázquez, *The analysis of coating flows in a strip*, J. Diff. Eqs., 121 (1995), 134–182.

[2] A. Friedman and J.J. Velázquez, *Time-dependent coating flows in a strip, Part I: The linearized problem*, Transactions Amer. Math. Soc., to appear.

[3] A. Friedman and J.J. Velázquez, *Liouville type theorems for fourth order elliptic equations in a half-space*, Transactions Amer. Math. Soc., to appear.

[4] A. Friedman, O Bruno and F. Reitich *Asymptotic behavior for a coalescence problem*, Trans. Amer. Math. Soc., 338 (1993), 133–158.

[5] T.H. Whitesides and D. Ross, *Experimental and theoretical analysis of the limited coalescence process: stepwise limited coalescence*, J. Colloid and Interface Science, 196 (1995), 48–59.

[6] A. Friedman and B. Hu, *Head-media interaction in magnetic recording*, Archive Rat. Mech. Anal., to appear.

[7] B. Morton and M. Elgersma, *A new computational algorithm for 7R spatial mechanisms*, Mech. Mach. Theory, 31 (1996), 24–43.

# Index

## IMA SUMMER PROGRAMS

1987 Robotics
1988 Signal Processing
1989 Robustness, Diagnostics, Computing and Graphics in Statistics
1990 Radar and Sonar (June 18 - June 29)
New Directions in Time Series Analysis (July 2 - July 27)
1991 Semiconductors
1992 Environmental Studies: Mathematical, Computational, and Statistical Analysis
1993 Modeling, Mesh Generation, and Adaptive Numerical Methods for Partial Differential Equations
1994 Molecular Biology
1995 Large Scale Optimizations with Applications to Inverse Problems, Optimal Control and Design, and Molecular and Structural Optimization
1996 Emerging Applications of Number Theory
1997 Statistics in Health Sciences.

## SPRINGER LECTURE NOTES FROM THE IMA:

*The Mathematics and Physics of Disordered Media*
Editors: Barry Hughes and Barry Ninham
(Lecture Notes in Math., Volume 1035, 1983)

*Orienting Polymers*
Editor: J.L. Ericksen
(Lecture Notes in Math., Volume 1063, 1984)

*New Perspectives in Thermodynamics*
Editor: James Serrin
(Springer-Verlag, 1986)

*Models of Economic Dynamics*
Editor: Hugo Sonnenschein
(Lecture Notes in Econ., Volume 264, 1986)

**The IMA Volumes in Mathematics and its Applications**

*Current Volumes:*

23 **Signal Processing Part II: Control Theory and Applications of Signal Processing** L. Auslander, F.A. Grünbaum, J.W. Helton, T. Kailath, P. Khargonekar, and S. Mitter (eds.)

24 **Mathematics in Industrial Problems, Part 2** A. Friedman

25 **Solitons in Physics, Mathematics, and Nonlinear Optics** P.J. Olver and D.H. Sattinger (eds.)

26 **Two Phase Flows and Waves** D.D. Joseph and D.G. Schaeffer (eds.)

27 **Nonlinear Evolution Equations that Change Type** B.L. Keyfitz and M. Shearer (eds.)

28 **Computer Aided Proofs in Analysis** K. Meyer and D. Schmidt (eds.)

29 **Multidimensional Hyperbolic Problems and Computations** A. Majda and J. Glimm (eds.)

30 **Microlocal Analysis and Nonlinear Waves** M. Beals, R. Melrose, and J. Rauch (eds.)

31 **Mathematics in Industrial Problems, Part 3** A. Friedman

32 **Radar and Sonar, Part I** R. Blahut, W. Miller, Jr., and C. Wilcox

33 **Directions in Robust Statistics and Diagnostics: Part I** W.A. Stahel and S. Weisberg (eds.)

34 **Directions in Robust Statistics and Diagnostics: Part II** W.A. Stahel and S. Weisberg (eds.)

35 **Dynamical Issues in Combustion Theory** P. Fife, A. Liñán, and F.A. Williams (eds.)

36 **Computing and Graphics in Statistics** A. Buja and P. Tukey (eds.)

37 **Patterns and Dynamics in Reactive Media** H. Swinney, G. Aris, and D. Aronson (eds.)

38 **Mathematics in Industrial Problems, Part 4** A. Friedman

39 **Radar and Sonar, Part II** F.A. Grünbaum, M. Bernfeld, and R.E. Blahut (eds.)

40 **Nonlinear Phenomena in Atmospheric and Oceanic Sciences** G.F. Carnevale and R.T. Pierrehumbert (eds.)

41 **Chaotic Processes in the Geological Sciences** D.A. Yuen (ed.)

42 **Partial Differential Equations with Minimal Smoothness and Applications** B. Dahlberg, E. Fabes, R. Fefferman, D. Jerison, C. Kenig, and J. Pipher (eds.)

43 **On the Evolution of Phase Boundaries** M.E. Gurtin and G.B. McFadden

44 **Twist Mappings and Their Applications** R. McGehee and K.R. Meyer (eds.)

45 **New Directions in Time Series Analysis, Part I** D. Brillinger, P. Caines, J. Geweke, E. Parzen, M. Rosenblatt, and M.S. Taqqu (eds.)

46 **New Directions in Time Series Analysis, Part II**
D. Brillinger, P. Caines, J. Geweke, E. Parzen, M. Rosenblatt, and M.S. Taqqu (eds.)

47 **Degenerate Diffusions**
W.-M. Ni, L.A. Peletier, and J.-L. Vazquez (eds.)

48 **Linear Algebra, Markov Chains, and Queueing Models**
C.D. Meyer and R.J. Plemmons (eds.)

49 **Mathematics in Industrial Problems, Part 5** A. Friedman

50 **Combinatorial and Graph-Theoretic Problems in Linear Algebra**
R.A. Brualdi, S. Friedland, and V. Klee (eds.)

51 **Statistical Thermodynamics and Differential Geometry of Microstructured Materials**
H.T. Davis and J.C.C. Nitsche (eds.)

52 **Shock Induced Transitions and Phase Structures in General Media** J.E. Dunn, R. Fosdick, and M. Slemrod (eds.)

53 **Variational and Free Boundary Problems**
A. Friedman and J. Spruck (eds.)

54 **Microstructure and Phase Transitions**
D. Kinderlehrer, R. James, M. Luskin, and J.L. Ericksen (eds.)

55 **Turbulence in Fluid Flows: A Dynamical Systems Approach**
G.R. Sell, C. Foias, and R. Temam (eds.)

56 **Graph Theory and Sparse Matrix Computation**
A. George, J.R. Gilbert, and J.W.H. Liu (eds.)

57 **Mathematics in Industrial Problems, Part 6** A. Friedman

58 **Semiconductors, Part I**
W.M. Coughran, Jr., J. Cole, P. Lloyd, and J. White (eds.)

59 **Semiconductors, Part II**
W.M. Coughran, Jr., J. Cole, P. Lloyd, and J. White (eds.)

60 **Recent Advances in Iterative Methods**
G. Golub, A. Greenbaum, and M. Luskin (eds.)

61 **Free Boundaries in Viscous Flows**
R.A. Brown and S.H. Davis (eds.)

62 **Linear Algebra for Control Theory**
P. Van Dooren and B. Wyman (eds.)

63 **Hamiltonian Dynamical Systems: History, Theory, and Applications**
H.S. Dumas, K.R. Meyer, and D.S. Schmidt (eds.)

64 **Systems and Control Theory for Power Systems**
J.H. Chow, P.V. Kokotovic, R.J. Thomas (eds.)

65 **Mathematical Finance**
M.H.A. Davis, D. Duffie, W.H. Fleming, and S.E. Shreve (eds.)

66 **Robust Control Theory** B.A. Francis and P.P. Khargonekar (eds.)

67 **Mathematics in Industrial Problems, Part 7** A. Friedman

68 **Flow Control** M.D. Gunzburger (ed.)

69 **Linear Algebra for Signal Processing**
A. Bojanczyk and G. Cybenko (eds.)

70 **Control and Optimal Design of Distributed Parameter Systems**
J.E. Lagnese, D.L. Russell, and L.W. White (eds.)

71 **Stochastic Networks** F.P. Kelly and R.J. Williams (eds.)

72 **Discrete Probability and Algorithms**
D. Aldous, P. Diaconis, J. Spencer, and J.M. Steele (eds.)

73 **Discrete Event Systems, Manufacturing Systems, and Communication Networks**
P.R. Kumar and P.P. Varaiya (eds.)

74 **Adaptive Control, Filtering, and Signal Processing**
K.J. Åström, G.C. Goodwin, and P.R. Kumar (eds.)

75 **Modeling, Mesh Generation, and Adaptive Numerical Methods for Partial Differential Equations** I. Babuska, J.E. Flaherty, W.D. Henshaw, J.E. Hopcroft, J.E. Oliger, and T. Tezduyar (eds.)

76 **Random Discrete Structures** D. Aldous and R. Pemantle (eds.)

77 **Nonlinear Stochastic PDEs: Hydrodynamic Limit and Burgers' Turbulence** T. Funaki and W.A. Woyczynski (eds.)

78 **Nonsmooth Analysis and Geometric Methods in Deterministic Optimal Control** B.S. Mordukhovich and H.J. Sussmann (eds.)

79 **Environmental Studies: Mathematical, Computational, and Statistical Analysis** M.F. Wheeler (ed.)

80 **Image Models (and their Speech Model Cousins)**
S.E. Levinson and L. Shepp (eds.)

81 **Genetic Mapping and DNA Sequencing**
T. Speed and M.S. Waterman (eds.)

82 **Mathematical Approaches to Biomolecular Structure and Dynamics**
J.P. Mesirov, K. Schulten, and D. Sumners (eds.)

83 **Mathematics in Industrial Problems, Part 8** A. Friedman

84 **Classical and Modern Branching Processes**
K.B. Athreya and P. Jagers (eds.)

85 **Stochastic Models in Geosystems**
S.A. Molchanov and W.A. Woyczynski (eds.)

86 **Computational Wave Propagation**
B. Engquist and G.A. Kriegsmann (eds.)

87 **Progress in Population Genetics and Human Evolution**
P. Donnelly and S. Tavare (eds.)

88 **Mathematics in Industrial Problems, Part 9** A. Friedman